INNOVATING CLIMATE GOVERNANCE

After the perceived failure of global approaches to tackling climate change, enthusiasm for local climate initiatives has blossomed worldwide, suggesting a more experimental approach to climate governance. *Innovating Climate Governance: Moving Beyond Experiments* looks critically at climate governance experimentation, focusing on how experimental outcomes become embedded in practices, rules and norms. Policy which encourages local action on climate change rather than global burden sharing suggests a radically different approach to tackling climate issues. This volume reflects on what climate governance experiments achieve, as well as what happens after and beyond these experiments. A bottom-up, polycentric approach is analysed, exploring the outcomes of climate experiments and how they can have broader, transformative effects in society. Contributions offer a wide range of approaches and cover more than fifty empirical cases internationally, making this an ideal resource for academics and practitioners involved in studying, developing and evaluating climate governance.

BRUNO TURNHEIM is Research Associate at the Department of Geography at King's College London, Research Fellow with the Manchester Institute of Innovation Research (MIoIR) at the University of Manchester and Associate Fellow with the Science Policy Research Unit (SPRU) at the University of Sussex. The overriding theme of his research concerns the role of innovation and transformative change for sustainability. His current research focuses on the governance of sustainability transitions, conceptual and methodological integration of research approaches to sustainability transitions, the role of local experimentation for transformative change, the diffusion of system innovation and the destabilisation of sociotechnical regimes.

PAULA KIVIMAA is Senior Research Fellow with the Science Policy Research Unit (SPRU) at the University of Sussex, Senior Researcher with the Finnish Environment Institute (SYKE), and Docent at Aalto University School of Business. Kivimaa, an expert in sustainability transition and innovation studies, focuses on the interface of policy and innovation. Recently, she has conducted novel research on policy mixes, intermediaries and experiments in transitions, covering a range of empirical contexts from transport to bioenergy and energy efficiency of buildings. Kivimaa frequently engages with policy-oriented audiences, including invited talks at the Organisation for Economic Co-operation and

Development (OECD) and the Finnish Ministries of the Environment and of Transport and Communications.

FRANS BERKHOUT is Executive Dean of the Faculty of Social Science and Public Policy and Professor of Environment, Society and Climate in the Department of Geography at King's College London. His work is concerned with science, technology, policy and sustainability, with a focus on climate change. He was a lead author in the fourth (2007) and fifth (2014) assessment reports of the Intergovernmental Panel on Climate Change (IPCC) and is Chair of UK Future Earth.

INNOVATING CLIMATE GOVERNANCE

Moving Beyond Experiments

Edited by

BRUNO TURNHEIM
King's College London

PAULA KIVIMAA
University of Sussex

FRANS BERKHOUT
King's College London

CAMBRIDGE
UNIVERSITY PRESS

CAMBRIDGE
UNIVERSITY PRESS

University Printing House, Cambridge CB2 8BS, United Kingdom

One Liberty Plaza, 20th Floor, New York, NY 10006, USA

477 Williamstown Road, Port Melbourne, VIC 3207, Australia

314–321, 3rd Floor, Plot 3, Splendor Forum, Jasola District Centre, New Delhi – 110025, India

79 Anson Road, #06-04/06, Singapore 079906

Cambridge University Press is part of the University of Cambridge.

It furthers the University's mission by disseminating knowledge in the pursuit of
education, learning, and research at the highest international levels of excellence.

www.cambridge.org
Information on this title: www.cambridge.org/9781108417457
DOI: 10.1017/9781108277679

First published 2018

Printed in the United Kingdom by Clays, St Ives plc

A catalogue record for this publication is available from the British Library.

Library of Congress Cataloging-in-Publication Data
Names: Turnheim, Bruno, 1984- editor. | Kivimaa, Paula, editor.
| Berkhout, F. (Frans), editor.
Title: Innovating climate governance : moving beyond experiments / edited by Bruno Turnheim
(King's College London), Paula Kivimaa (University of Sussex), and Frans Berkhout
(King's College London).
Description: Cambridge : Cambridge University Press, [2018] | Includes bibliographical
references and index.
Identifiers: LCCN 2017055332 | ISBN 9781108417457 (hardback : alk. paper)
Subjects: LCSH: Climate change mitigation. | Community development.
Classification: LCC TD171.75 .I56 2018 | DDC 363.738/7456–dc23
LC record available at https://lccn.loc.gov/2017055332

ISBN 978-1-108-41745-7 Hardback

Contents

Contributors

Harro van Asselt is Professor of Climate Law and Policy at University of Eastern Finland Law School, Joensuu, Finland, and Senior Research Fellow at the Stockholm Environment Institute, Oxford, United Kingdom.

Frans Berkhout is Executive Dean, Faculty of Social Science and Public Policy and Professor of Environment, Society and Climate at King's College London, United Kingdom.

Joannette J. Bos is Senior Lecturer at the Monash Sustainability Development Institute, Monash University, Australia.

Bas Breman is Senior Researcher at Alterra, Wageningen University and Research, the Netherlands.

Harriet Bulkeley is Professor of Geography at Durham University, United Kingdom.

Arwin van Buuren is Professor of Public Administration at Erasmus University Rotterdam, the Netherlands.

Luís Carvalho is Senior Researcher at the Centre of Studies in Geography and Spatial Planning at the University of Porto, Portugal.

Vanesa Castán Broto is Professorial Fellow at the Department of Geography and the Urban Institute at the University of Sheffield, United Kingdom.

Gerald Jan Ellen is Senior Researcher Delta Governance at Deltares, the Netherlands.

Megan Farrelly is Senior Lecturer at the School of Social Sciences at Monash University, Australia.

Niki Frantzeskaki is Associate Professor at the Dutch Research Institute for Transitions (DRIFT) at Erasmus University Rotterdam, the Netherlands.

Eva Heiskanen is Professor at the Consumer Society Research Centre at the University of Helsinki, Finland.

Katharina Hölscher is PhD Researcher at the Dutch Research Institute for Transitions (DRIFT) at Erasmus University Rotterdam, the Netherlands.

Michael Howlett is Burnaby Mountain Chair in the Department of Political Science at Simon Fraser University, British Columbia, Canada.

Dave Huitema is Professor of Environmental Policy at the Netherlands Open University and the Vrije Universiteit Amsterdam, the Netherlands.

Andrew Jordan is Professor of Environmental Policy at the Tyndall Centre for Climate Change Research at the University of East Anglia, Norwich, United Kingdom.

Andrew Karvonen is Assistant Professor in Urban and Regional Studies at the KTH Royal Institute of Technology, Stockholm, Sweden.

Paula Kivimaa is Senior Research Fellow at the Science of Policy Research (SPRU) at the University of Sussex, United Kingdom, and Senior Researcher at the Finnish Environment Institute (SYKE), Helsinki, Finland.

Irina Lazzerini is Sustainable Energy Expert at the Energy Community, Vienna, Austria.

Corniel van Leeuwen is Researcher at GovernEUR at Erasmus University Rotterdam, the Netherlands.

Derk Loorbach is Professor and Director at the Dutch Research Institute for Transitions (DRIFT) at Erasmus University Rotterdam, the Netherlands.

Kaisa Matschoss is Senior Researcher at the Consumer Society Research Centre at the University of Helsinki, Finland.

Sreeja Nair is Postdoctoral Research Fellow in the School of Social Sciences at Nanyang Technological University, Singapore.

Helen Pallett is a Lecturer in the Science, Society and Sustainability research group in the School of Environmental Sciences at the University of East Anglia, Norwich, United Kingdom.

Jitske van Popering-Verkerk is Postdoctoral Researcher in the Department of Public Administration at Erasmus University Rotterdam, the Netherlands.

Bruno Turnheim is Research Associate at the Department of Geography at King's College London, United Kingdom; Associate Faculty at the Science of Policy Research at the University of Sussex, United Kingdom; and Research Fellow at the Sustainable Consumption Institute at the University of Manchester, United Kingdom.

Heleen Vreugdenhil is Researcher at Deltares and at the Technology, Policy and Management Group at Delft University of Technology, the Netherlands.

Preface

This volume is a product of our collaboration within the EU-COST Network on Innovation in Climate Governance (INOGOV) that aims to contribute to understanding how climate governance can be innovated by exploring the relationships between sources, patterns and effects of climate policy and governance. The network brings together a research community which informed and acted as a test bed for our reflections on the topic of experimentation in climate governance. A network like INOGOV is an ideal space for the exploration of new ideas, as it brings together a group of committed and intellectually diverse experts, generating opportunities for critical reflection and cross-fertilisation of perspectives. A workshop organised at the Finnish Environment Institute (SYKE) by Mikael Hildén on experimentation in climate governance in March 2015 convinced us not only of the richness of scholarship on the topic but also of the need to combine an emerging understanding of the process of experimentation with an understanding of the legacy and broader outcomes of experiments: 'beyond experiments'. How is it that, despite an explosion in the number of local and non-state experiments with climate governance, very little is known about how and whether these initiatives may add up, learn from each other, persist through time, travel or become institutionalised? These are the questions we sought to explore.

This book was further shaped by the Centre for Innovation and Energy Demand (CIED), which provided an enabling research environment for two of us during the writing period. In particular, the Centre's focus on experimentation and accelerated diffusion allowed us to actively draw on parallels between energy innovation and climate governance innovation.

With funding from INOGOV and CIED, we followed up with our own workshop in Brighton, at the Science Policy Research Unit (SPRU), University of Sussex, on 25–27 April 2016. The event attracted about twenty scholars from eight countries to discuss questions related to how experiments in climate governance can benefit society, particularly in terms of lasting impacts, beyond the (spatial and temporal) boundaries for which they have been originally designed. This volume brings together many of the contributions to this workshop, providing an overview of the diversity of scholarship on this topic. The level of engagement of participants during and after the workshop, their commitment to constructive debate and critical feedback and their willingness to learn from neighbouring fields made our task as editors very enjoyable. We wish to thank all contributors and hope to have accurately conveyed the relevance and substance of their ideas in our introducing and concluding chapters.

We are also very grateful to Andy Jordan and Dave Huitema (Chairs of INOGOV) for supporting the idea early on and for encouragement and advice, as well as for contributing a chapter (led by Harro van Asselt) that offers a crucial overview of the changed opportunities and challenges for climate experimentation post-Paris Agreement. We wish to thank Johanna Forster (Tyndall Centre, University of East Anglia) for her generous support to enabling our workshop. We are indebted to Benjamin Sovacool (SPRU) for continued advice and for hosting us in Brighton.

This project would not have been feasible without the financial support from INOGOV (COST funded Action No. IS1309) and the Centre for Innovation and Energy Demand (CIED) funded by the RCUK EUED Programme (grant EP/KO11790/1). The work has also been supported by funding from the Strategic Research Council at the Academy of Finland (decision number 293405) and from the green.eu project (building the inno4sd network), which has received funding from the European Union's Horizon 2020 research and innovation programme under grant agreement No. 641974.

1

Beyond Experiments

Innovation in Climate Governance

BRUNO TURNHEIM, PAULA KIVIMAA AND FRANS BERKHOUT

1.1 Experiments in Climate Governance

In this edited volume, we are interested in understanding how experiments in climate governance can lead to broader changes in rules, practices, norms and other wider outcomes of efforts to respond to the challenges of climate change. We start with three observations about experiments in climate governance as entry points for some more general reflections about how change in governance comes about *from below*, rather than as a result of coordinated policymaking from above.

First, climate governance experimentation has become a legitimate object of research and is a practice attracting interest among policymakers and citizens. There is, therefore, an opportunity to analyse the motivations, direct outputs and broader outcomes of these initiatives. We want to ask what do these experiments add up to, and whether they influence deeper change in the legitimacy and effectiveness of climate governance. This is the broader policy context on this volume.

Second, climate governance experimentation is linked to the search for new ways of dealing with the causes and consequences of climate change, often at the margins of formal and established governance regimes, and in ways that are often temporary and local. We seek to understand what happens beyond this initial experimental setting. How do the ideas, networks and capabilities that emerge and are partially stabilised in experimental settings come to have a broader impact across policy and political systems? This provides a general problem and intellectual challenge for this volume.

Third, climate governance experimentation is a multifaceted object of study that compels a view from different perspectives. With this volume we seek to draw on the richness of a variety of conceptual and methodological traditions to further our understanding of governance experimentation in the context of climate change. In particular, we have sought to bring scholars of governance and of innovation together to reflect on climate policy experiments and their broader impacts beyond the original experimental setting. We do so by encouraging and setting the terms of a constructive dialogue between these quite distinct approaches. This provides an interdisciplinary orientation to this volume.

1.1.1 The Growing Attention to Climate Governance Experiments

In common with other areas of policy studies (Greenberg, Linksz and Mandell, 2003; Tassey, 2014), there has been a growing academic and policy interest in experimentation in governing the causes and consequences of climate change over recent years. This is evident across different scales, from local communities and cities (cf. Blok and Tschötschel, 2016) to policy communities (McFadgen and Huitema, 2017) and international organisations.

There appear to be a number of reasons for this revived interest (Sabel and Zeitlin, 2012). First, experimentation is seen as a mode of response well suited to the challenges of mitigating climate change and adapting to climate risks. It is argued that experiments are better attuned to the complex, situated and uncertain character of the climate change problem than more traditional modes of governing through national and international policy (e.g. Broto and Bulkeley, 2013; Bulkeley, Broto and Edwards, 2014; McFadgen and Huitema, 2016). In particular, governance experiments appear to be fitting when responding to uncertainties and incentive problems confronted by local climate action. The literatures on risk governance (Renn, Klinke and van Asselt, 2011), polycentric governance (Ostrom, 2010; Jordan et al., 2015) and urban experimentation (Bulkeley et al., 2014) recognise the limited capacity of national and international policy regimes to address global climate change effectively. This failure accounts for the 'groundswell of actions on climate change mitigation and adaptation from cities, regions, businesses, and civil society organizations' (Chan et al., 2015:476). According to some commentators, the failure of the 15th session of the Conference of the Parties (COP 15) in Copenhagen (2009) strengthened a mandate for decentralised, bottom-up climate interventions – a shift in climate governance internationally that was confirmed at COP21 in Paris (2016) that placed greater emphasis on voluntarism at the national level ('pledge and review') and a greater role for non-state action and subnational actors (van Asselt, Huitema and Jordan, Chapter 2). The 'experimental turn' in climate governance can be viewed as a rejection of the perceived failures of coordinated and global approaches to climate action, whether that coordination was achieved through governments or markets. Experimentalism has been presented as an entrepreneurial approach, stressing agency over coordination, with coordination itself viewed as emergent and *organic*, drawing on the norms, incentives and relationships of actors at a more granular level.

Second, experimentalism is being embraced as a principle for action in an area that is fraught with uncertainty, complexity, diffuse authority and agency, justified by the need to design provisional goals and to fine-tune through comparative learning (Sabel and Zeitlin, 2010; De Búrca, Keohane and Sabel, 2014). In this view, experimentation is more than a means to an end. The function of experimentation is not merely to encourage learning or to build up actor coalitions that can propel change. Instead, experimentalism is seen as a new approach to climate governance itself; that is, it is a transformation in governance in its own right. This debate on experimentalist governance extends well beyond the issue of climate change (Sabel and Zeitlin, 2012), but it points to a deeper set of problems in complex, polycentric and multilevel governance systems.

In either case, experimentation represents a challenge to climate governance as conventionally conceived and practiced. Often, experiments are inscribed in processual narratives linking demonstrations, pilots and field trials with the promise of a deeper link to motivations and incentives of actors, and generalisable and replicable approaches. However, the true value of governance experiments in serving as microcosms that can be disseminated and reproduced is in question. For example, it is not clear how experimentation can generate outcomes beyond learning by those directly engaged in them, and the body of evidence documenting successful replication remains thin (Kivimaa et al., 2017).

Current enthusiasm for experimentation in climate governance explains the proliferation of initiatives and schemes. It also creates increased scope for reflection about the goals and consequences of experimentation: what experiments may lead to, beyond their particular and bounded contexts, and whether they can influence changes in norms, incentives, rules, behaviours and relationships more generally. This volume seeks to explore the question of what lies after and *beyond* experiments. In doing so, we aim to contribute to a critical analysis of climate governance experimentation. If experiments are largely uncoordinated and entrepreneurial initiatives by new coalitions of actors, what direct outputs do the experiments produce and how do they come to have broader influence? What notions of diffusion, reproduction and embedding can best describe the process by which the multiple possible outputs of experiments come to generate broader outcomes? These are deep conceptual challenges which each of the contributions in this collection grapple with and which we return to in Chapter 12.

1.1.2 Framing the Problem: Embedding Climate Governance Experiments

A good starting point for a volume about climate governance experimentation is to understand how experimentation became a promising approach for addressing global climate change. Climate change has been labelled a 'wicked problem *par excellence*' (Dryzek, 1987; Jordan et al., 2010; Levin et al., 2012; Jordan and Huitema, 2014c). This is because of the inherent messiness, uncertainty and intractability of climate change, and the complexities of incentives and resistance to possible responses, whether through the mitigation of climate-forcing emissions or adaptation to the impacts of climate change. There is no simple 'climate fix'. Instead a range of activities have been taken, for example, in the domains of renewable energy (Baker and Sovacool, 2017), low carbon mobility (Hopkins and Highham, 2006) and building energy demand reduction (Kivimaa and Martisikai-nen, 2017) with the hope of partly alleviating the problems of climate change. Awareness and knowledge of climate change is partial and contested, and incentives for action may be weak and perverse. The nature of climate change and the difficulties it poses for collective decision-making and coordination (with a global commons, blurred and differentiated responsibilities, asymmetries in costs and benefits of action, and so on) have precipitated a general search for novel forms of governance that are more exploratory, flexible and multivalent (Biermann et al., 2012; Burch et al., 2014; Hale and Roger, 2014; Jordan and Huitema, 2014a; Chan et al., 2015). Global state-led climate governance has been characterised by, for some, a disappointing record and a history of political impasses (e.g. Levin et al., 2012; Kanie et al., 2012). This record has played a role in energising the search for new ways of handling the causes and implications of climate change.

The search for and analysis of innovative forms of climate governance has been a feature of academic commentary over the past decade (Jordan and Huitema, 2014a, 2014b, 2014c; Upham et al., 2014). This includes the crafting of new governance arrangements, as well as analysis of how new modes and instruments of governance are implemented and evaluated (Huitema et al., 2011). Jordan and Huitema (2014a, 2014b) describe policy and governance innovation as significant novelty linked to the emergence of a new policy, its diffusion and effects. Part of this debate has concerned the role of experiments in generating innovations in governance, including a variety of attempts at defining climate governance experiments. Kivimaa et al. (2017:2) argue that governance experiments 'can either constitute (deliberate) interventions that aim at solving problems or developing new practices (as in pilots or demonstration projects), or they are conducted in order to learn about the effects of (limited) interventions for future (more large-scale) interventions'. Experiments can embody governance innovation but present the additional 'opportunity to tinker with new approaches, practices or institutions on a small scale and/or temporarily' (Kivimaa et al., 2017:2). It has also been argued that experimentation is less directed than innovation – often associated with the adoption of an idea in a market – and is therefore more open-ended and oriented towards exploration (e.g. Schot, Kanger and Verbong, 2016). This approach is also used in the definition of an urban sustainability experiment developed by Sengers et al. (2016:21): 'An inclusive, practice-based and challenge-led initiative designed to promote system innovation through social learning under conditions of deep uncertainty and ambiguity.'

The literature on climate governance (Hoffmann, 2011) has been interested in exploring novel forms of action 'beyond, below and outside the state-dominated climate regime' (Jordan and Huitema, 2014c). However, the analyses are often narrowly focussed on the realm of policy itself, with little consideration for the social, institutional and material aspects of governance (Bulkeley et al., 2014). Conversely, the literature on socio-technical experiments in the context of sustainability transitions (Kemp, Rip and Schot, 2001; Berkhout et al., 2010; Smith and Raven, 2012; Späth and Rohracher, 2012) has been less concerned with specific applications in policy and governance (Kivimaa et al., 2017). This gap represents a serious constraint on the broader outcomes potentially generated by experiments in governance for sustainability. Experimental initiatives tend to be situated in time and place, operate in relative isolation, and may require further refinement and consolidation to become impactful more widely. Beyond the talk of the need to *scale up*, there is little insight into how the direct outputs of experiments can be reproduced and embedded to achieve significant impact on climate change problems.

We believe that a useful next step is to define ways to harness learning from experiments with new instruments, modes and approaches to climate governance, and at the same time consider critically the shortcomings of experimentation as a solution to the wicked problem of climate change. This may be done by studying the careers of individual climate experiments and experimental practices, examining the variety of climate action on the ground, and theorising and tracking their broader outcomes on the way climate governance is done and what effects this may have at different scales of analysis.

Taking the notion of climate governance experimentation seriously, this volume focuses on the career, relevance and adequateness of climate governance experiments beyond their experimental nature, and beyond their own institutional contexts. It explores the expansion, reproduction and embedding of climate governance experiments as they turn into more than experiments.

1.1.3 Approach: Interdisciplinarity and Empirical Variety

With this volume, we have sought to capture a wide range of perspectives on climate governance experiments, reflecting the diversity of approaches proposed in the literature and in practice. It brings together contributions from a range of approaches to climate governance experiments – governance understood in the broadest sense as forms of coordination of state and society toward collective interest (Pierre and Peters, 2005) – and to experimentation and sustainability innovation more generally. Rather than advocating for a particular view, we seek to provide a broad picture of existing concepts, representing a variety of approaches, the different challenges they identify and the main strategies they offer for governing climate change. We also seek to reflect critically on current interest in experimentation, which is far from a benign and neutral term. We find it particularly useful to mobilise rich empirical cases to support this critical line of enquiry.

We have sought to stimulate a constructive dialogue between the different approaches critically engaging with experimentation for climate governance. We have convened contributions by climate governance and innovation scholars, understood widely as studying the introduction of novelty to sociotechnical systems and the institutional and material reconfigurations that may ensue. When doing so, it became evident that other related fields are also relevant in approaching the central questions posed by 'beyond experiments', and this volume therefore also builds from selected approaches in science and technology studies, geography, and policy studies. This has resulted in contributions that together span a wide variety of concepts and analytical frames, providing different lenses through which to appreciate the challenges and lasting impacts of climate governance experimentation. We hope to have contributed to mapping out the contours of this intellectual space and the multiple opportunities it offers.

Rather than providing an overarching framework, our aim has been to make sense of rich and varied new directions for research, guiding contributions into a coherent direction, so as to explore the scope for cross-fertilisation. This has led us to offer a general problem framing (climate governance, experiments, embedding), a set of concepts to the problem at hand and what we see as underlying master processes (articulation and alignment at the level of systems) that each contribution deals with in specific ways. This allows us to explore a variety of current analytical contributions, unpack their significance and identify their potential complementarities. We also explore how these different conceptual frames may be 'bridged' (Turnheim et al., 2015).

Climate governance experimentation is rapidly evolving, presenting challenges to practitioners and researchers. For this reason, we thought it relevant to seek out a variety of interesting and novel empirical cases, focussing on their richness and diversity (see further Section 1.3). Contributions to this volume critically engage with real-world cases of climate governance experimentation, further supporting our collective exploration with empirical context and contributing to our broader conceptual ambitions. Besides obvious benefits in terms of generating inductive insights on the conduct of climate governance experiments in practice, this allows contributions to produce greater clarity about the phenomena at hand: experimentation and embedding.

1.2 Conceptual Starting Points

In this section, we start mapping out the main analytical challenges of this volume in greater detail and apply a variety of concepts to make sense of the emergent significance of experimentation in the climate governance. We recognise an inherent problem with providing strict definitions, mainly because conceptual flexibility is valuable when engaging with an emerging problem area characterised empirically by a multiplicity of entry points and because our background aim is to bring together contributions from a range of perspectives, themselves often invoking varying and incompatible conceptual tools. After more than a year of convening and mediating interdisciplinary conversations on the topic, we see our task as clarifying the range of perspectives and where they may be bridged. This implies mapping, unpacking and exposing the variety of useful perspectives, rather than reaching conceptual closure. We do so around a clear intellectual programme, which concerns understandings of experimentations, their emergence and consolidation into new 'orders' and the different ways in which they become embedded in practices, institutions and regimes of governance.

1.2.1 Experimentation and Experiments

Whereas in natural and engineering science, as well as some fields of social science like psychology and economics, the experiment is a methodological framework for testing knowledge claims against well-established criteria of significance, the notion of experimentation which we use here is significantly different. In the context of governance, experimentation is associated with more open-ended initiatives usually designed to test the feasibility or effectiveness of a novel governance practice in which emergent or unexpected outcomes may be the anticipated product. Although there is likely to be an evaluation framework for governance experiments, the process and criteria for evaluation are expected to be flexible to some extent, needing to take account of the unfolding and emergent nature of the impacts which may be observed. Typically, experiments will be expected to lead to changes, whether these relate to the pursuit of new knowledge, new practices, new solutions, or the enrolment of new actors (see Karvonen, Chapter 11; Pallett, Chapter 5). As in natural science experiments, scepticism is important to the success of governance experiments since, in practice, experiments may be mobilised to make up for the lack of more systemic action (Howlett, 2014), which can also lead to 'reframed policy innovations' (Upham et al., 2014). As background for the contributions to this collection, we outline a number of ways in which experimentation has been framed in existing literature, highlighting also what we see as the main focus of innovation studies and governance studies.

1.2.2 Why Experiment? An Overview of Metaphors

One aim of this collection is to explore the different ways in which climate governance experiments are conceived in social science. It is these 'creation myths' associated with experiments which will serve as the template for ideas about the broader outcomes of experiments on policy and governance. Here, we review briefly some of the main metaphors which have been employed in talking about experiments in governance and innovation studies.

Experimentation as Method: Testing Hypotheses

The term 'experimentation' originates from scientific method and experimental practice in laboratory contexts (see Pallett, Chapter 5). In this original form, experimentation is often inscribed within a positivist understanding of knowledge production through a primarily deductive logic, and a general understanding that a hypothesis can be formulated and then 'tested'. In that context, experiments are seen as allowing for the testing of hypotheses through repeatable observations and the introduction of variations in a controlled setting (the laboratory). Strict controlled environments do not exist in the social realm and, hence, call for methodological adjustments in the context of climate governance (e.g. 'uncontrolled experiments', 'field experiments'). An experimental approach carries with it the illusion of control over an environment, the social world and its complexity. The notion of laboratory has been transposed into the social world, in settings such as living labs (Veeckman et al., 2013), where strategic experimentation is taking place, requiring the creation of contained and to some degree 'controlled' spaces (Evans, 2011). Spatial and temporal bounding become central concerns (e.g. Karvonen, Chapter 11).

Experimentation as Testing: Selecting Designs that Work

Related to the preceding discussion, and against the background of classical understandings of innovation, experimentation is often seen as the initial step (e.g. 'from theory to practice', 'from design to implementation' or 'from idea to market'). Here experimentation is seen as a means for selecting promising designs and specifying challenges on which to focus for further development. A novel idea is trialled so as to establish its feasibility, identify potential problems and guide further adjustments. This view is tied to an understanding of experimentation as a source of strategic learning to be exploited.

Experimentation is here seen as the more or less systematic testing of ideas. Within business innovation, these can be referred to as 'trial-and-error problem-solving processes and strategies for experimentation used in the development of new products and services' (Thomke, von Hippel and Franke, 1998:315). This form of experimentation typically involves a simplified version of an innovative product or service and may go through a series of stages. Pilots seek to test for feasibility and acceptability, while demonstration projects aim to refine further the performance potential of an innovation (Hoogma, 2000). For highly regulated products, like pharmaceuticals, safety and efficacy testing is part of the demonstration phase. Experimenting as testing informs the notion of policy piloting, where learning can occur in a specific setting before wider deployment (see Nair and Howlett, Chapter 9; van Buuren et al., Chapter 8), or to more symbolically display leadership on a particular issue.

Experimentation as Transformational Strategy: Learning by Doing

Beyond the limits of scientific method and hypothesis testing, experiments are generally associated with the acquisition of new skills and knowledge. In such an understanding, experimentation may refer to trial-and-error learning. Learning by doing is also explicit in most definitions of experiments (Smith, 2006; Berkhout et al., 2010). Experimentation produces specific kinds of interventions, observations and inferences that may be strategically mobilised for governance purposes (Pahl-Wostl, 2009). The key aspect of experimentation becomes a process of recursive learning, which is seen as enabling improvement through iterative cycles of designing, making and adjusting (see Farrelly and Bos, Chapter 6; Karvonen, Chapter 11). Experimentation can, in this view, also be seen as a specific disposition of individuals or organisations, to be resilient under turbulent environments and is linked to notions of improvisation and organisational adaptation (Tushman and Romanelli, 1985; Weick, 1998). Therefore, learning happens both during and after experiments, on the basis of individual projects and at a more aggregate level. From our perspective of 'beyond experiments', learning after an experiment also appears important. In this category, higher order learning has been described as a measure of success (Brown and Vergragt, 2008) that manifests itself through, for example, changed discourses and practices, as well as policy and institutional change resulting from experimentation (Kivimaa et al., 2017).

Experimentation as Radical Novelty Creation: Opening Up Alternatives

A related metaphor understands experimentation as a source of novelty. On the one hand, such novelty can consist in relatively small variations from existing processes, offering scope for incremental improvement. On the other, radical innovation can be seen as novelty creation well beyond the boundaries of existing frameworks (of knowing, of doing, of thinking, etc.). Such a view is closely associated with an understanding that radical change tends to come from outside the prevailing ways of doing things and involves breaking conventions by experimentalists. Experimentation can then be seen as thinking beyond existing paradigms to solve previously intractable problems, or to chart new possibilities. In a policy and governance context, this may involve seeing problems under, above and between existing jurisdictions (Jordan and Huitema, 2014c).

Experimentation as Nurturing: Fostering Alternatives in Protected Spaces

Linked to the innovation metaphor is an emphasis on the fragility and lack of 'fitness' of any form of novelty. Mokyr (1990) referred to path-breaking innovations as 'hopeful monsters' that have yet to fulfil their potential and may carry a number of intrinsic problems. From this comes the idea that experiments are organised for nurturing and protecting early and vulnerable seeds of change. Experiments are seen as small-scale initiatives in the earliest stages of innovation processes that do not yet conform to existing socio-technical contexts (Schot, Hoogma and Elzen, 1994; Berkhout et al., 2010). Due to their inherent fragility, new socio-technical configurations can be strategically nurtured in 'niches' (cf. Kemp et al., 2001), understood as 'protected spaces' where external selection pressures cannot exert their full influence (Smith and Raven, 2012). Within this evolutionary understanding of change, experimentation is seen as an activity enabling a variety of options and solutions to be generated and their relevance explored. This view sees a role for experimentation in an understanding of transformative change that originates from and grows in innovation niches, and eventually may break through to challenge (and overtake) an established regime.

Experimentation as Politics: Performing Reality

Experimentation is not a value-free proposition. On the contrary, engaging with the world through experimentation is reminiscent of the generalisation of a scientific method to all realms of society – in our case climate governance. However open or narrow the transposition of the laboratory metaphor to the social realm, experimentation has a performative dimension with deep implications. Experimentation implies the appreciation and acceptance of a worldview and a set of tools, mobilised to produce collective realities (see Castán Broto and Bulkeley, Chapter 4). In short, experimentation can be seen as a process of ordering the socio-material world. The experimental process and its concrete outputs, by articulating and establishing a certain kind of reality, define what is important and worth observing, make predictions about broader outcomes and seek to validate these through actions. An experimental attitude contributes to 'governing' the perceptions and actions of individual actors and decision-makers by, for instance, favouring certain approaches over others, legitimising certain forms of epistemic authority and permitting and preventing access of certain actors. In this sense, experimentation can be seen as 'politics by other means'. This becomes salient when knowledge generated by experiments is mobilised to support or justify specific political decisions. Looking at experimentation from this perspective enables a critical and reflexive appreciation of experimental settings, with attention to the normative, political and cultural implications of climate governance experiments. It may open up to alternative ways of handling experimentation beyond disciplined, authoritative and exclusive procedures for knowledge generation. This view of experimentation can also be empowering insofar as it supports the possibility for reordering collective realities.

This overview of understandings and metaphors of experimentation provides a range of narratives for our exploration of experimentation in climate governance. Experiments are associated with hopeful promises, carry a novelty value, are inherently fragile, tentative and temporary, are often mobilised instrumentally and also denote a certain creative capacity to envision alternative ways of seeing, doing and valuing. Experimentation is often inscribed in a wider frame of innovation – whether this is motivated by a discourse of 'betterment', a logic of continuous adaptation to changing realities or less instrumental perspectives on change. Proposing experimentation as a legitimate approach to governance calls into question the motives, supporting narratives and expectations, as well as models of change being invoked. It may be wise to remain on your guard when scholars start speaking about experiments.

1.2.3 A Dialogue between Studies of Governance and Innovation Studies

A number of insights have informed the design of this collection. First, the observation that while research on climate governance has increasingly come to frame a concern around the *innovation* of new modes and instruments of governance (Jordan and Huitema, 2014b), it has not engaged in any deep way with the perspectives that have emerged over the past fifty years or so in economics, sociology and history on the role of new knowledge and technology in the economy and society more broadly. Second, the observation that contemporary innovation studies concerned with *socio-technical* change often take an unnuanced view of policy and governance and appear to have had little impact on the practice of policy and governance. There appears to be room to learn from scholars of government and governance. Third, both traditions have shown a marked interest in experiments, whether these are experiments in governance (Sabel and Zeitlin, 2012; McFadgen and Huitema, 2016) or experiments (or niches) as sources of radical novelty leading to wider socio-technical change (Kemp, Schot and Hoogma, 1998; Seyfang and Smith, 2007; Smith and Raven, 2012; Seyfang et al., 2014). In both traditions there has been an interest in the role of 'outsider' and radical actors in generating new ideas and ways of doing things, and how these activities can be nurtured and protected so that they can prove their value over the longer term.

Finally, both traditions have not been interested in experiments in their own right, but primarily as a source of alternatives which may come to challenge and overturn dominant paradigms and practices. For both governance and innovation studies, the interest in experiments has therefore ultimately been about how the concrete outputs of experiments may generate broader impact on the governance or socio-technical systems which they address. This may be through the adoption of ideas, through the translation of skills and capabilities or through a deeper reordering of norms and institutional rules and arrangements. So it is clear that the two traditions have much to say to each other. Climate governance studies may be enriched by insights from innovation studies, and the flow of ideas in the other direction may be just as fruitful. Here we provide a sketch of the main analytical frames mobilised in innovation studies and governance studies, suggesting how they may contribute to the research agenda, and what the potential for cross-fertilisation is.

Innovation Studies

Innovation studies is concerned with the role of science, technology and innovation in economies and societies and draws on a range of disciplines including economics, history, science and technology studies, sociology and cultural studies. Classical work on innovation (Freeman, Clark and Soete, 1982; Freeman and Louçã, 2001; Fagerberg, 2004) distinguishes between radical technical change through sustained search and experimentation with alternatives and incremental change through trial-and-error processes and learning-by-doing in which tacit and practical knowledge are built up about new ways of doing things (van de Ven et al., 1999).[1] Emphasis is placed on the combination of resources (knowledge, skills, networks and material resources) with sufficient freedom and openness to create opportunities for novelty, while balancing the need for continuity and stability in production and consumption systems.

Risk, uncertainty, surprises and failures are viewed as intrinsic to innovative processes and explain the behaviour of actors involved in these processes, whether they are innovators or adopters of innovations. This points to the risks innovation poses for established actors – whether these are firms, public organisations or societal organisations – and explains why experiments often originate from newcomers and marginal actors. This relates to deeper debates about

---

[1] These trial-and-error processes may also be regarded as experimental, leading to a bifurcation in the understandings of experimentation in innovation studies: first as organised processes of disruptive or radical innovation, and second as a generalised process leading to incremental innovation.

innovation which we can roughly trace as a progression away from linear models (from science to R&D to innovation) emphasising the development of technological knowledge, towards a greater interest in innovation within socio-technical systems in which norms, rules and social relationships are always embedded in and shaped by material technologies (Geels, 2002, 2005; Smith, Stirling and Berkhout, 2005). This new wave of innovation studies, which is interested in larger-scale system innovation over the long run, also takes a broader view of socio-technical systems as 'configurations that work' (Rip and Kemp, 1998).

Classical studies of innovation diffusion draw heavily on the work of Rogers (1962) who viewed diffusion as a communication process between members of a social system, leading to the development of an 'epidemic model' of adoption and diffusion. Rogers and subsequent contributions (Von Hippel, Thomke and Sonnack, 1999) argued that adoption includes a measure of reinvention or further adaptation and the 'fitting' of innovations to user needs and practices. In this sense, adoption and diffusion must be seen as an integral part of the innovation process. Nevertheless, it can be argued that innovation studies generally have an under-theorised approach to processes of adoption, diffusion and transferability of innovations, whether technical or organisational.

Socio-technical innovation studies argue that experimentation, as a distributed and problem-oriented search process, requires enabling environments, systems and institutions (Hekkert et al., 2007; Smith, Voss and Grin, 2010). Strategic Niche Management (Kemp et al., 1998; Kemp et al., 2001), for instance, argues that 'protected spaces' are needed for experimental activities to develop into more stable configurations (Smith and Raven, 2012) to be able to overturn incumbent socio-technical regimes. Communication alone does not explain the processes of adoption and adaptation that occur as new ways of doing things are diffused more broadly. Instead the enabling role of collective learning and knowledge development, new networks and alliances that convene disparate interests, and the definition of shared goals, norms and expectations are significant in bringing about a reconfiguration of a broader socio-technical system, hence 'system innovation'.

Governance and Policy Studies

Dating back to the 1950s (Hoppe, Coenen and van den Berg, 2016), governance and policy studies can broadly be defined to contain the disciplines of political science, public administration, economics and legal studies, among others. Policy studies are focused on the content, processes and effects of government policies in their surrounding social and political contexts (Hoppe et al., 2016) while governance studies are broader, with interest in the governing activities of social, economic, political and administrative actors (Kooiman, 2003) and articulations of policy, politics and polity (Lange et al., 2013). As with innovation studies, governance studies draw from multiple theories and literature to study structures, mechanisms and policies in place, as well as the processes by which they unfold and influence real or potential impacts. The latter is particularly addressed in studies of policy evaluation and appraisal – that have not considered the evaluation of policy innovations or experiments explicitly.

Governance and policy studies have come to experimentation with an interest in *responses*, *solutions* and *arrangements* that do not readily fit pre-existing policy channels (e.g. Sabel and Zeitlin, 2010), or which emerge in response to fragmented political orders and authorities, as is the case with 'climate experiments' (Hoffmann, 2011). There is a recognition of the limitation of traditional models of evidence-based governance in the face of 'wicked' problems and their associated uncertainties and incentive problems.

Two main views seem to co-exist and overlap. On the one hand, experimentation is seen as a structured process of search initiated by institutionally situated actors. 'Policy experiments' are seen as pilots, typically upstream from a process that may lead to policy innovation. This transfer is often unsuccessful because of the difficulty to 'scale up' or 'diffuse' new ideas and arrangements (Vreugdenhil, Taljaard and Slinger, 2012). Such difficulties may arise from a lack of willingness to address more fundamental problems (Howlett, 2014). Experiments may also significantly challenge established ways of doing and evaluating policy (Martin and Sanderson, 1999).

On the other hand, experimentation may be seen as a means for 'shaking up' governance arrangements around new narratives, logics, interests, incentive structures and evaluation schemes, as witnessed in recent enthusiasm about the role of bottom-up, voluntary and entrepreneurial non-state climate action (Jordan et al., 2015), and in the research focus on climate experiments (Abbott, 2012; Bulkeley et al., 2012). Climate governance experiments have been framed as alternate means of responding or attending to climate change (Hoffmann, 2011), contrasting with the more formal governance regimes of international agreements, with targets and coordination mechanisms codified in law. These local climate governance experiments may provide space and mandates for new actors (e.g. the city) to lead on generating situated solutions to collective action problems, tapping into grassroots energy and ingenuity. Within this latter view, experimentation has become associated with emerging centres of authority and governance that are inherently more distributed and networked (polycentrism) and with alternative ways of achieving coordinated action that are more tentative, emergent and self-organising and that avoid the costs and rigidities of centralised coordination.

1.2.4 Beyond Experiments: What Are the Outputs of Experiments and What Influence Do They Have?

We want to understand experiments in terms of their wider impacts once the experimental phase is over and to explore experimentation as a practice for generating changes in climate governance. For this reason, we seek to understand the *becoming* of climate governance experiments, in search of what are their outputs that may be communicated, reproduced and embedded beyond particular experiments themselves, thus, generating broader outcomes. We are interested in the processes by which experiments may become relevant from a systemic perspective.

We also believe that a focus on 'beyond experiments' enables us to articulate a number of important tensions of relevance to experimentation as a way forward to address climate governance challenges. Our interest in 'beyond' – preposition, adverb and noun – is tied to its potential to generate critical analysis. In this respect we see 'beyond' as taking up different meanings:

- **A temporal** dimension which can be summarised as 'after'. Along this dimension, we are interested in the fate of the concrete outputs of experiments (ideas, norms, people and ways of doing things), starting from the well-documented observation that governance experiments are usually short-lived, often abandoned as political priorities change or a cycle of funding ends, as are their legacies. So, we ask whether experiments can be sustained through time. This relates to the 'longevity' of governance experiments as well as to deeper changes that may emerge from experimentation processes. In this second temporal sense, we are interested in the broader outcomes and implications of experiments, whether this is seen as learning, conceptual or hypothesis testing, network and alliance formation, the opening of new development paths and so on. We also link this back to a question about stability and change in innovation studies. The temporal dimension justifies a need for more critical analyses of the long-term life and effects of climate governance experimentation.
- **A spatial** dimension which articulates the tension between the situated nature of individual experiments and expectations about more generic and transferable outcomes of experimentation. This relates to the questions about whether experiments and their concrete outputs can be sustained across different scales and spaces. If we look beyond individual experiments, we see a variety of processes by which the outputs of experiments can become mobile, affecting their contributions to transformative change.
- **An evaluative** dimension which links back to the pragmatic justification for experimentation as a new means of governing (experimental governance). Questions here concern common evaluation criteria (e.g. relevance,

effectiveness, efficiency and coherence), as well as new questions related to scaling or mainstreaming or the cumulative effects of experimentation (or, in other words, how experimental governance initiatives may 'add up' and substantially contribute to the long-term mitigation of climate change). We believe this evaluative dimension to also be crucial in terms of critically addressing the 'success' in experimentation, and the extent to which 'failure' may be equally (if not more) valuable in terms of learning from experimentation – a common theme in innovation studies. We would like to confront the idea that experimentation always leads to broader positive outcomes by, for instance, attending to the proponents of experimentation in climate governance and their motivations.

In short, we believe that a satisfactory account of experiments would deal with the temporal and spatial boundaries of an experiment, beyond which it has broader influence, and with the criteria that are being used to judge and value that influence.

1.2.5 Diffusion of Experimental Outputs as a Process of Embedding

In searching for a single concept to capture the wide variety of ways in which the outputs of experiments may have a wider influence, we have chosen to employ the concept of embedding. We search for a concept of mobility, expansion and diffusion of experimental outputs that combines a notion of transfer and exchange with a conception of the broader institutional, political and normative settings that may be transformed by these transfers and which, in turn, respond to such exchange. We want to develop an idea of embedding that responds to the core notion of a governance system as a configuration of rules, responsibilities, values and outcomes that works, and which to some extent comes to be reworked or reconfigured as a result of its response to the outputs of experiments. Outcomes are not unidirectional, but always a derivative of a further interplay of new and existing ideas and practices. We hope to demonstrate with this volume that experiments become relevant if and as they engage with the challenge of their embedding – something that is often lacking in practice but for which we can already derive some guiding insights and lessons.

Embedding as Object of Research: Why Focus on Embedding?

We argue that despite a growing interest in experimentation in general and for climate governance more specifically, comparatively less attention has been paid to governance *after and beyond* the scope of experiments. Focussing on the embedding of experiments enables us to capture a specific 'moment' and 'site' for innovation in society: the acquisition of momentum and wider relevance of individual initiatives. Given that experimentation is increasingly seen as a solution to overcome lock-in and path dependencies that prevent effective climate change mitigation and its governance, we need to further understand the actual outputs of climate experiments, their implications and the processes by which experimental outputs become embedded and create more lasting outcomes in society.

Broadening the Understanding of Embedding

Theories of embeddedness take their rooting in the field of economic history (Polanyi, 2001) and economic sociology (Granovetter, 1985) as a way to problematise the influence of wider social environments on, in these circumstances, economic activity. In this context, economic activity is understood as 'embedded' in a wider set of social institutions and relationships. Long-term economic change goes hand in hand with concomitant institutional change and their mutual embedding. Generalising from this view, embeddedness can be applied to any sphere of social exchange as a way of understanding its relationship to the wider social, institutional and cognitive environments in which they operate. Embedding and embeddedness have taken up a significant importance in innovation studies, where 'societal embedding' is seen as a process of mutual shaping and adjustment between innovation and its wider context (Leonard-Barton, 1988; Nye, 1998; Rip and Kemp, 1998; Boschma, Lambooy and Schutjens, 2002).

Embeddedness can be thought of as a form of interdependence and connectivity between activities, structures and contexts. In the frame of system innovation, embedding can be seen as a process of articulation and alignment of experiments in conjunction with wider institutional or cultural settings. This can be operationalised at different levels and across multiple dimensions.

Embedding Climate Governance Experimentation

The embedding of climate governance experiments captures a process by which experiments become more than experiments, for example, by encouraging the wider uptake of a new approach, by informing the establishment of new forms of governance, by entering mainstream discourses and practices, by prefiguring wider transformation, by challenging established ways of doing and so on. What we mean by *embedding* is a process by which governance experiments develop or influence beyond the initial context within which a new way of doing things has been configured, involving a recasting of its scope and enrolling new agents into the project of performing an alternative mode or method of governing, and through such a process transforming climate governance itself. Not all experiments become embedded, and it is important to pay more attention to the extent to which experiments meet their claim in achieving the impacts they are intended or envisioned to achieve.

Given this broad understanding of embedding within the context of climate change governance, a number of views about the main processes involved can be distinguished. These differ in terms of the core mechanisms, temporalities, scales, institutional dynamics and agents of change that they mobilise. The contributions to this volume offer a range of different perspectives of embedding, and underlying models of change, which we summarise and discuss in the concluding chapter to this volume.

1.3 Contributions to This Volume

1.3.1 Overview

Chapters in this volume provide an indication of the breadth of research concerned with the role of experiments in climate governance and their contribution to climate change mitigation and adaptation. They provide an impression of the variety of climate governance experimentation on the ground, with particular emphasis on non-state local action, and the different ways in which this is contributing to and challenging state-led climate governance practices. While an edited volume can only cover a part of this expanding domain, the chapters cover substantial empirical territory (see Table 1.1): a wide range of cases in several countries and spanning all the major continents. They show a balance between concerns of climate adaptation (responding to the impacts of changing climate and variability) and mitigation (reducing emissions of greenhouse gases to the atmosphere). Many chapters illustrate examples of experimentation and beyond, predominantly in the energy and water management domains, but also in buildings, urban regeneration, agriculture and participatory knowledge production.

Given the dominance of urban issues in recent research on climate governance experimentation (evidenced by the growing influence of networks such as ICLEI and C40 on climate issues), it is not surprising that many chapters place emphasis on local urban and municipal scales. However, the volume also looks at national and international governance regimes and covers interactions across different levels of governance.

In conceptual terms, the contributions span a broad spectrum of ways of thinking about experiments, their concrete outputs and their broader outcomes, including the establishment of experimentation as a mainstream logic for governance. In this way, the collection of chapters meets our goal to examine a range of takes on what can happen after an experimental stage. In our efforts to consider productive interactions between governance studies and innovation studies to examine this topic, we also mobilised contributions from neighbouring fields, including urban and economic geography and transitions studies. In addition, some

Table 1.1. *Comparative overview of empirical book contributions (Chapters 3–11)*

#	Cases — Authors	Initiatives	Country	Domain	IPCC — Adaptation	Mitigation	Scale — Municipal	Urban	National	Multilevel	Time frame	Literature — Governance	Innovation	Transitions	Geography	Other	Lead actor — Policy	Civil society	Research	Private
3	Carvalho and Lazzerini (PT, IT)	Multiple in 6 states	US	Electricity		•	•			•	1997–2016	•	•		•		•	•		
4	Castán Broto and Bulkeley (UK)	8	IN, MEX, US, ZA, DE, BR, HK, UK	Multiple	•	•	•	•				•	•		•	STS	•	•	•	•
5	Pallett (UK)	Multiple	UK	Multiple (science)	n/a	n/a			•	•	2000–today		•			STS		•		
6	Farrelly and Bos (AUS)	Multiple	AUS	Buildings, water	•	•	•			•	1990s–today	•	•	•			•			
7	Hölscher et al (NL)	Multiple	NL	Energy, water mgmt.	•	•		•		•	2005–today	•	•	•	•			•	•	•
8	Van Buuren et al (NL)	11 pilots	NL, DE	Water mgmt.	•		•		•				•				•			
9	Nair and Howlett (SG, US)	14	IN	Agriculture	•	•	•		•	•	1990–today		•			Policy analysis	•			
10	Heiskanen and Matschoss (FI)	2	FI	Multiple		•		•					•	•	Evaluation					
11	Karvonen (UK)	Multiple	n/a	Unspecified	•			•	•			•	•		•		•	•		

of the chapters touch on more 'specialised' theoretical perspectives. They build on influences derived from Science and Technology Studies (STS), policy evaluation, field theory, resilience theory, participation and democratic theory, assemblage and urban metabolism. The contributions in this volume bear witness to the range of relevant theories and approaches to studying climate governance experiments and their embedding (see Table 1.1).

In terms of the empirical contributions, the chapters present illustrative cases of experiments and their embedding within a relatively recent time frame, focussing on the 1990s onwards, with most cases covering the past ten to fifteen years. This temporal focus reflects the strong momentum of climate experimentation in the field in recent years. Similarly, while representing all world regions, a significant proportion of the evidence discussed in this volume is drawn from Northern European countries. This focus reflects an empirical reality, as climate governance experimentation remains an emerging phenomenon shaped by public discourses in that part of the world – with, for instance, the Netherlands and Finland playing a particularly visible role in institutionalising this experimental stance.

1.3.2 Structure of This Volume

We have chosen to organise this volume in two parts, reflecting broad motivations for climate governance experimentation, namely *exploration* and *transformation*. Because these are inherently intertwined, most contributions are relevant to both aspects. Nonetheless, the contributions tend to emphasise one over the other, which reflects the inherent dilemma that their joint pursuit involves in practice.

Part I Experiments: Exploring Innovations in Climate Governance

Climate governance experiments have the potential to open up new ways of attending to and living with the challenges of climate change. Part I focuses on the conduct of climate governance experiments, how and where experimentation takes place, who participates, the various formats of experimentation, its exploratory nature and the kinds of direct and indirect outputs generated. It does so by mobilising rich case studies and multi-case comparisons of climate governance experimentation in practice. We are interested in better understanding the range of broader outcomes of experiments that extend beyond their original setting, and how they can contribute to the crafting of more durable alternatives. Within this frame, the kinds of knowledge and experiences arising from climate change experiments as they become exploitable, transferable or scalable beyond their original application context are explored. This leads us to focus on mechanisms supporting the shift from situated climate governance experiments to more generic and mobile alternatives that can gain traction in a variety of contexts. Beyond experiments are new territories for their embedding.

Part II Beyond Experiments: Transforming Climate Governance

Climate governance experiments, beyond enabling exploration and the search for alternative ways of doing, can have profound effects on governance arrangements themselves and the socio-technical systems in which they are embedded. Part II focuses on the transformative outcomes of governance experiments, as their proliferation challenges and reconfigures governance structures and processes in place. This leads us to focus on how climate governance experiments are taken up in existing climate governance regimes and how climate experiments transform governance and other milieus through which they pass. Embracing climate experimentation as a new *modus operandi* calls for new (e)valuations of success and reinventing the craft of climate governance and decision-making altogether. We ask what climate governance may look like if it becomes reconfigured around the opportunities and promises deriving from the handling of experimentation. Beyond experiments are unchartered territories for climate governance.

1.4 Conclusion

This volume has been designed to provide an exploration of the range of conceptual approaches to the embedding of experiments leading to innovations in climate governance, each foregrounding specific patterns, mechanisms, roles, strategies, tensions and opportunities. By presenting a variety of perspectives, we hope to enrich an understanding of the possible avenues for climate governance experimentation. This kind of novel cross-fertilisation between perspectives can generate important lessons for understanding the dynamics of innovation in climate governance arrangements and experiments in particular, with the aim of contributing to effective strategies for addressing the serious societal risks associated with global climate change.

We bring together a number of rich and original empirical contributions, in-depth case studies and more comparative approaches. Together, we hope that they convey the variety of contexts in which climate governance experimentation is being pursued, contribute to the understanding of how experiments are becoming embedded across a variety of contexts and, thus, feed into both experimental designs where embedding is regarded from the outset and institutional designs that encourage and make room for the outcomes of experiments.

As the search for new governance responses to climate change accelerates, we are likely to see more experiments. So we need to be equipped: remain critical as to motivations, recognise an increasingly distributed and polycentric governance landscape, introduce clarity of objectives and evaluation and develop pathways to learning and competence building. We want a situated analysis of the mobility and influence or a variety of possible outputs of experiments and the challenge they pose to sense-making, positionalities, values and the constitution of orders. We want to explore experiments as a precursor to generalising ways of doing things and the potential of experimentalism as a governance approach in its own right.

Clarifying and mobilising the idea of embedding experiments is a major conceptual challenge. There are different understandings of this process. Conventionally these have included theories of diffusion and the idea of 'upscaling'. Diffusion has been formalised in a variety of ways in innovation and governance studies, while upscaling has been much less well theorised. We do not prescribe one over others. However, we do want to develop an argument that argues that innovation – developing a new ways of doing things – requires a process of rule-making and practical *embedding* that includes both the modification of the new, as well as the reconfiguration of existing systems. Embedding is, therefore, a mutual process of adaptation of novelty and of the governance context within which innovations flowing from governance experiments become embedded. This process will follow context-specific and historical patterns.

References

Abbott, K. W. (2012). The transnational regime complex for climate change. *Environment and Planning C: Government and Policy*, 30(4), 571–590.

Baker, L., and Sovacool, B. K. (2017). The political economy of technological capabilities and global production networks in South Africa's wind and solar photovoltaic (PV) industries. *Political Geography*, 60, 1–12.

Berkhout, F., Verbong, G., Wieczorek, A. J., Raven, R., Lebel, L., and Bai, X. (2010). Sustainability experiments in Asia: Innovations shaping alternative development pathways? *Environmental Science & Policy*, 13(4), 261–271.

Biermann, F., Abbot, K., Andresen, S., Backstrand, K., Bernstein, S., et al. (2012). Transforming governance and institutions for global sustainability: Key insights from the Earth System Governance Project. *Current Opinion in Environmental Sustainability*, 4(1), 51–60.

Blok, A., and Tschötschel, R. (2016). World port cities as cosmopolitan risk community: Mapping urban climate policy experiments in Europe and East Asia. *Environment and Planning C: Government and Policy*, 34(4), 717–736.

Boschma, R. A., Lambooy, J. G., and Schutjens, V. (2002). Embeddedness and innovation. In Tayloer, M., and Leonard, S. (eds.), *Embedded Enterprise and Social Capital*. Aldershot: Ashgate, 19–37.

Broto, V. C., and Bulkeley, H. (2013). A survey of urban climate change experiments in 100 cities. *Global Environmental Change*, 23(1), 92–102.

Brown, H. S., and Vergragt, P. J. (2008). Bounded socio-technical experiments as agents of systemic change: The case of a zero-energy residential building. *Technological Forecasting and Social Change*, 75(1), 107–130.

Bulkeley, H., Andonova, L., Bäckstrand, K., Betsill, M., Compagnon, D., et al. (2012). Governing climate change transnationally: Assessing the evidence from a database of sixty initiatives. *Environment and Planning C: Government and Policy*, 30(4), 591–612.

Bulkeley, H. A., Broto, V. C., and Edwards, G. A. S. (2014). *An Urban Politics of Climate Change: Experimentation and the Governing of Socio-technical Transitions*. London: Routledge.

De Búrca, G., Keohane, R. O., and Sabel, C. (2014). Global experimentalist governance. *British Journal of Political Science*, 44(3), 477–486.

Burch, S., Shaw, A., Dale, A., and Robinson, J. (2014). Triggering transformative change: A development path approach to climate change response in communities. *Climate Policy*, 14(4), 467–487. doi: 10.1080/14693062.2014.876342.

Chan, S., van Asselt, H., Hale, T., Abbott, K. W., Beisheim, M., et al. (2015). Reinvigorating international climate policy: A comprehensive framework for effective nonstate action. *Global Policy*, 6(4), 466–473.

Dryzek, J. S. (1987). *Rational Ecology: Environment and Political Economy*. New York: Basil Blackwell.

Evans, J. P. (2011). Resilience, ecology and adaptation in the experimental city. *Transactions of the Institute of British Geographers*, 36(2), 223–237.

Fagerberg, J. (2004). Innovation: A guide to the literature. In. J. Fagerberg, D. C. Mowery, and R. R. Nelson (eds.), *The Oxford Handbook of Innovation* (pp. 1–26). Oxford: Oxford University Press.

Freeman, C., Clark, J., and Soete, L. (1982). *Unemployment and Technical Innovation: A Study of Long Waves and Economic Development*. London: Pinter.

Freeman, C., and Louça, F. (2001). *As Time Goes By*. Oxford: Oxford University Press.

Geels, F. W. (2002). Technological transitions as evolutionary reconfiguration processes: A multi-level perspective and a case-study. *Research Policy*, 31, 1257–1274. doi: 10.1016/S0048-7333(02)00062-8.

Geels, F. W. (2005). Processes and patterns in transitions and system innovations: Refining the co-evolutionary multi-level perspective. *Technological Forecasting and Social Change*, 72(6), 681–696. doi: 10.1016/j.techfore.2004.08.014.

Granovetter, M. (1985). Economic action and social structure: The problem of embeddedness. *American Journal of Sociology*, 91(3), 481–510.

Greenberg, D. H., Linksz, D., and Mandell, M. (2003). *Social Experimentation and Public Policymaking*. Washington, DC: The Urban Institute Press.

Hale, T., and Roger, C. (2014). Orchestration and transnational climate governance. *Review of International Organizations*, 9(1), 59–82.

Hekkert, M. P., Suurs, R., Negro, S., Huhlmann, S., and Smits, R. (2007). Functions of innovation systems: A new approach for analysing technological change. *Technological Forecasting and Social Change*, 74(4), 413–432.

Von Hippel, E., Thomke, S., and Sonnack, M. (1999). Creating breakthroughs at 3M. *Harvard Business Review*, 77, 47–57.

Hoffmann, M. J. (2011). *Climate Governance at the Crossroads: Experimenting with a Global Response after Kyoto*. Oxford: Oxford University Press.

Hoogma, R. (2000). *Exploiting Technological Niches: Strategies for Experimental Introduction of Electric Vehicles*. Enschede: Twente University Press.

Hopkins, D., and Higham J. (eds.). (2016). *Low Carbon Mobility Transitions*. Oxford: Goodfellow Publishers.

Hoppe, T., Coenen, F., and van den Berg, M. (2016). Illustrating the use of concepts from the discipline of policy studies in energy research: An explorative literature review. *Energy Research and Social Science*, 21, 12–32.

Howlett, M. (2014). Why are policy innovations rare and so often negative? Blame avoidance and problem denial in climate change policy-making. *Global Environmental Change*, 29, 395–403. doi: 10.1016/j.gloenvcha.2013.12.009.

Huitema, D., Jordan, A., Massey, E., Rayner, T., van Asselt, H., et al. (2011). The evaluation of climate policy: Theory and emerging practice in Europe. *Policy Sciences*, 44(2), 179–198.

Jordan, A., and Huitema, D. (2014a). Innovations in climate policy: Conclusions and new directions. *Environmental Politics*, 23(5), 906–925. doi: 10.1080/09644016 .2014.924209.

Jordan, A., and Huitema, D. (2014b). Innovations in climate policy: The politics of invention, diffusion, and evaluation. *Environmental Politics*, 23(5), 906–925.

Jordan, A., and Huitema, D. (2014c). Policy innovation in a changing climate: Sources, patterns and effects. *Global Environmental Change*, 29, 387–394.

Jordan, A., Huitema, D., van Asselt, H., Rayner, T., and Berkhout, F. (2010). *Climate Change Policy in the European Union: Confronting the Dilemmas of Mitigation and Adaptation?* Cambridge: Cambridge University Press.

Jordan, A. J., Huitema, D., Hildren, M., van Asselt, H., Rayner, T. J., et al. (2015). Emergence of polycentric climate governance and its future prospects. *Nature Climate Change*, 5(11), 977–982. doi: 10.1038/nclimate2725.

Kanie, N., Betsill, M. M., Zondervan, R., Biermann, F., Young, O. R. (2012). A charter moment: Restructuring governance for sustainability. *Public Administration and Development*, 32(3), 292–304.

Kemp, R., Schot, J., and Hoogma, R. (1998). Regime shifts to sustainability through processes of niche formation: The approach of strategic niche management. *Technology Analysis & Strategic Management*, 10(2), 175–198.

Kemp, R. P. M., Rip, A., and Schot, J. W. (2001). Constructing transition paths through the management of niches. In Garud, R., and Karnoe, P. (eds.), *Path Dependence and Creation*. Mahwah, NJ: Lawrence Erlbaum Associates, 269–299.

Kivimaa, P., Hildén, M., Huitema, D., Jordan, A., and Newig, J. (2017). Experiments in climate governance: A systematic review of research on energy and built environment transitions. *Journal of Cleaner Production*, 169: 17–29.

Kivimaa, P., and Martiskainen, M. (2017). Innovation, low energy buildings and intermediaries in Europe: systematic case study review. *Energy Efficiency*, 1–21.

Kooiman, J. (2003). *Governing as Governance*. London: Sage.

Lange, P., Driessen, P. P. J., Sauer, A., Bornemann, B., and Burger, P. (2013). Governing towards sustainability: Conceptualizing modes of governance. *Journal of Environmental Policy and Planning*, 15(3), 403–425.

Leonard-Barton, D. (1988). Implementation as mutual adaptation of technology and organization. *Research Policy*, 17(5), 251–267. doi: 10.1016/0048–7333(88)90006-6.

Levin, K., Cashore, B., Bernstein, S., and Auld, G. (2012). Overcoming the tragedy of super wicked problems: Constraining our future selves to ameliorate global climate change. *Policy Sciences*, 45(2), 123–152. doi: 10.1007/s11077-012-9151-0.

Martin, S., and Sanderson, I. (1999). Evaluating public policy experiment: Measuring outcomes, monitoring processes or managing pilots. *Evaluation*, 5(3), 245–258.

McFadgen, B., and Huitema, D. (2017). Are all experiments created equal? A framework for analysis of the learning potential of policy experiments in environmental governance. *Journal of Environmental Planning and Management*, 20(10) 1–20.

McFadgen, B., and Huitema, D. (2017). Experimentation at the interface of science and policy: A multi-case analysis of how policy experiments influence political decision-makers. *Policy Sciences*, 1–27. doi: 10.1007/s11077-017-9276-2

Mokyr, J. (1990). *The Lever of Riches: Technological Creativity and Economic Progress*. New York: Oxford University Press.

Nye, D. E. (1998). *Consuming Power: A Social History of American Energies*. Cambridge, MA: MIT Press.

Ostrom, E. (2010). Beyond markets and states: Polycentric governance of complex economic systems. *American Economic Review*, 100(3), 641–672. doi: 10.1257/aer.100.3.641.

Pahl-Wostl, C. (2009). A conceptual framework for analysing adaptive capacity and multi-level learning processes in resource governance regimes. *Global Environmental Change*, 19(3), 354–365.

Pierre, J., and Peters, B. (2005). *Governing Complex Societies: Trajectories and Scenarios*. Basingstoke and New York: Palgrave Macmillan.

Polanyi, K. (2001). *The Great Transformation: The Political and Economic Origins of our Time*. London: Beacon.

Renn, O., Klinke, A., and van Asselt, M. (2011). Coping with complexity, uncertainty and ambiguity in risk governance: A synthesis. *Ambio*, 40(2), 231–246.

Rip, A., and Kemp, R. (1998). Technological change. In Rayner, S., and Malone, L. (eds.), *Human Choice and Climate Change,* Vol. 2*, Resources and Technology*. Washington, DC: Batelle Press, 327–399.

Rogers, E. M. (1962). *Diffusion of Innovations*. New York: Free Press of Glencoe.

Sabel, C. F., and Zeitlin, J. (2010). Learning from difference: The new architecture of experimentalist governance in the EU. In Sabel, C. F., and Zeitlin, J. (eds.), *Experimentalist Governance in the European Union: Towards a New Architecture*. Oxford: Oxford University Press, 1–28.

Sabel, C. F., and Zeitlin, J. (2012). Experimentalist governance. In Levi-Faur, D. (ed.), *The Oxford Handbook of Governance*. Oxford: Oxford University Press.

Schot, J., Hoogma, R., and Elzen, B. (1994). Strategies for shifting technological systems: The case of the automobile system. *Futures*, 26(10), 1060–1076.

Schot, J., Kanger, L., and Verbong, G. (2016). The roles of users in shaping transitions to new energy systems. *Nature Energy*, 1(5), 16054. doi: 10.1038/nenergy.2016.54.

Sengers, F., Berkhout, F., Wieczorek, A. J., and Raven, R. (2016). Experimenting in the city: Unpacking notions of experimentation for sustainability. In Evans, J., Karvonen, A., and Raven, R. (eds.), *The Experimental City*. London and New York: Routledge, 15–31.

Seyfang, G., Hielscher, S., Hargreaves, T., Martiskainen, M., Smith, A. (2014). A grassroots sustainable energy niche? Reflections on community energy in the UK. *Environmental Innovation and Societal Transitions*, 13, 21–44. doi: 10.1016/j.eist.2014.04.004.

Seyfang, G., and Smith, A. (2007). Grassroots innovations for sustainable development: Towards a new research and policy agenda. *Environmental Politics*, 4016(4), 37–41. doi: 10.1080/09644010701419121.

Smith, A. (2006). Green niches in sustainable development: The case of organic food in the United Kingdom. *Environment and Planning C: Government and Policy*, 24(3), 439–458. doi: 10.1068/c0514j.

Smith, A., and Raven, R. (2012). What is protective space? Reconsidering niches in transitions to sustainability. *Research Policy*, 41(6), 1025–1036. doi: 10.1016/j.respol.2011.12.012.

Smith, A., Stirling, A., and Berkhout, F. (2005). The governance of sustainable socio-technical transitions. *Research Policy*, 34(10), 1491–1510. doi: 10.1016/j.respol.2005.07.005.

Smith, A., Voss, J. P., and Grin, J. (2010). Innovation studies and sustainability transitions: The allure of the multi-level perspective and its challenges. *Research Policy*, 39(4), 435–448. doi: 10.1016/j.respol.2010.01.023.

Späth, P., and Rohracher, H. (2012). Local demonstrations for global transitions: Dynamics across governance levels fostering socio-technical regime change towards sustainability. *European Planning Studies*, 20(3), 461–479.

Tassey, G. (2014). Innovation in innovation policy management: The Experimental Technology Incentives Program and the policy experiment. *Science and Public Policy*, 41(4), 419–424.

Thomke, S., von Hippel, E., and Franke, R. (1998). Modes of experimentation: An innovation process – and competitive – variable. *Research Policy*, 27(3), 315–332. doi: 10.1016/S0048-7333(98)00041-9.

Turnheim, B., Berkhout, F., Geels, F., Hof, A., McMeekin, A., Nykvist, B., and van Vuuren, D. (2015). Evaluating sustainability transitions pathways: Bridging analytical approaches to address governance challenges. *Global Environmental Change*, 35, 239–253.

Tushman, M. L., and Romanelli, E. (1985). Organizational evolution: Interactions between external and emergent processes and strategic choice. *Research in Organizational Behavior*, 8, 171–222.

Upham, P., Kivimaa, P., Mickwitz, P., and Åstrand, K. (2014). Climate policy innovation: A sociotechnical transitions perspective. *Environmental Politics*, 23(5), 774–794.

Veeckman, C., Schuurman, D., Leminen, S., and Westerlund, M. (2013). Linking living lab characteristics and their outcomes: Towards a conceptual framework. *Technology Innovation Management Review*, 3(12), 6–15.

van de Ven, A. H., Polley, D., Garud, R., and Venkataraman, S. (1999). *The Innovation Journey*. New York: Oxford University Press New York.

Vreugdenhil, H., Taljaard, S., and Slinger, J. H. (2012). Pilot projects and their diffusion: A case study of integrated coastal management in South Africa. *International Journal of Sustainable Development*, 15(1–2), 148–172.

Weick, K. E. (1998). Introductory essay: Improvisation as a mindset for organizational analysis. *Organization Science*, 9(5), 543–555.

2

Global Climate Governance after Paris

Setting the Stage for Experimentation?

HARRO VAN ASSELT, DAVE HUITEMA AND ANDREW JORDAN

2.1 Introduction

In a year in which the world experienced the highest temperature ever recorded (Carrington, 2017), the Paris Agreement became the fastest multilateral environmental agreement to formally enter into force. That Agreement marks a substantial shift in the course of global climate governance. It represents a move away from the approach followed under the Kyoto Protocol in which industrialised countries exclusively assumed a range of economy-wide and legally binding greenhouse gas targets for specific time periods. By contrast, the Paris Agreement applies to *all* – and not just developed – countries. In addition, rather than multilaterally agreeing on goals for each country, it allows countries to set their *own ambition levels* in the form of non-legally binding 'nationally determined contributions' (NDCs), which are subsequently made subject to review.

Whether the Agreement's 'pledge-and-review' approach will be sufficient to meet international climate objectives still remains to be seen, but it is clear that the pledges made by governments thus far are insufficient to have a realistic chance to meet the Agreement's goal of staying well below 2°C (Rogelj et al., 2016; UNEP, 2016), let alone the Agreement's new goal of pursuing efforts to stay below 1.5°C. Through its combination of nationally determined pledges and legally binding procedural obligations, the Paris Agreement represents a novel approach that is yet to be tried out. In that sense, it constitutes an experiment in multilateral cooperation – one with potentially very high stakes (Doelle, 2016).

But it is not just the Agreement's pledge-and-review approach that is the experiment: the broader approach to climate governance it signifies (and seeks to encourage) is also experimental, since a broader spectrum of approaches can now be tried out by a much wider array of parties, and because their eventual outcomes are to be systematically assessed. The Paris climate summit, as well as the decision adopting the Agreement (UNFCCC, 2015), saw a greater recognition and encouragement of the wealth of climate action by a variety of non-state and subnational actors all trying out their own approaches, including local governments, businesses and investors, civil society, etc. (Chan, Brandi and Bauer, 2016; Hale, 2016). In doing so, Parties made first steps in linking the upsurge of non-state and subnational action to the multilateral framework for state-based action. The greater involvement of new actors may offer new hope for meeting international climate objectives (Hsu et al., 2015; UNEP, 2016). However, as noted by Turn-heim, Kivimaa and Berkhout (Chapter 1), it remains to be seen whether these experiments by non-state and subnational actors will deliver transformational change.

In this chapter, we trace the evolution of global climate governance (Section 2.2). Specifically, we examine whether and to what extent we can observe an experimentalist turn in global climate governance, focusing on four elements of global experimentalist governance outlined by De Búrca, Keohane and Sabel (2013; 2014). These are: (1) the setting of framework goals and establishment of metrics in an open process involving the participation of a broad group of stakeholders; (2) decentralised implementation; (3) monitoring and reporting, including peer review; and (4) evaluation and revision of the goals in light of experiences. In doing so, we aim to shed light on the extent to which the global climate governance architecture post-Paris resembles the ideal type of global experimentalist governance, potentially setting the stage for experimentation at lower levels of governance (Section 2.3). In our conclusions, we sketch several challenges in ensuring the functioning of global experimentalist climate governance and offer some critical reflections on the concept (Section 2.4).

2.2 From Rio to Paris and Beyond: A Short History of Global Climate Governance

Twenty-five years in the making, the international regime established by the United Nations Framework Convention on Climate Change (UNFCCC) can be considered the world's leading response to the climate problem. The Convention was one of the major outcomes of the United Nations Conference on Environment and Development in Rio de Janeiro in 1992. At a time when scientific insights into the problem, including the precise human contribution to it, were still emerging, nations opted to set a broad goal to 'stabilize greenhouse gas emissions at a level that would avoid dangerous anthropogenic interference with the climate system' (Article 2, UNFCCC). Aside from a vaguely formulated and non-legally binding goal for developed countries to return their greenhouse gas emissions to 1990 levels by 2000 (Article 4.2[a]), the Convention did not specify its overall objective any further. It did, however, require parties to regularly report on their emissions and measures taken to reduce them and established the necessary institutional infrastructure to facilitate the functioning of the agreement.

The absence of concrete targets was an omission that parties sought to correct quickly, and indeed, through the adoption of the Kyoto Protocol in 1997, quantified emission limitation and reduction obligations were adopted for the group of developed countries. Those countries were offered flexibility to achieve their emission reduction targets by employing a host of market-based mechanisms. Furthermore, to ensure that the countries lived up to their legally binding obligations, a compliance mechanism was put in place that could avail of both carrots and sticks (Brunnée, 2001).

The years following the adoption of the Protocol proved to be rather turbulent. Parties struggled to reach agreement on the detailed rulebook for Kyoto implementation in 2000, leading to the breakdown of the Conference of the Parties (COP) in The Hague that year. This was followed by the announcement one year later that the world's largest emitter at the time, the United States, would withdraw from the Protocol, arguing that the treaty's exclusion of emerging economies (notably China) from emissions targets would put it at a competitive disadvantage. Following the United States' departure, parties managed to reach agreement on the rulebook in Marrakech in 2001, though this required concessions that led to a watering down of some of the rules. Agreement on the rulebook facilitated the participation of a number of key emitters (e.g. Australia and Japan), and – following a deal between the European Union (EU) with Russia – the Protocol finally entered into force in 2005.

Although parties began with implementing the Protocol ahead of the first commitment period (2008–2012), disagreements among parties continued to plague the climate regime. The major bone of contention became what to do in the period after 2012, when the Protocol's targets would expire. Particularly, some developed countries were adamant that international climate policy in the post–2012 period needed to include some form of mitigation action for emerging economies, such as China (which overtook the United States as the world's largest emitter in 2007) and India. Developing countries, conversely, resisted this push and preferred to stick to Kyoto's developed/developing country dichotomy (Rajamani, 2013). The tensions were unresolved by the time the COP in Copenhagen in 2009 commenced. At that conference, parties were supposed to adopt a successor agreement, but the combination of diverging positions of key countries and the chaotic conference organisation spelled failure, and the resulting Copenhagen Accord was not adopted by consensus (Dimitrov, 2010).

Just one year later, however, the 2010 Cancún Agreements confirmed the main elements of the Copenhagen Accord (UNFCCC, 2010). The Accord asked countries to make emission reduction pledges, which would subsequently be subject to a system of review. The Cancún Agreements anchored this system of 'pledge and review' by elaborating the reporting and review requirements for both developed and developing countries. As a consequence, both groups of countries were required to report more regularly, and in more detail (Gupta and van Asselt, 2017).

In parallel to these developments within the international climate regime, new governance initiatives started to emerge outside of it. For instance, governments created a range of new public-private partnerships promoting climate change technologies (van Asselt, 2014), such as the now defunct Asia-Pacific Partnership on Clean Development and Climate, and the Major Economies Meeting (continued by US President Obama as the Major Economies Forum on Energy and Climate in 2009). Although the United States drove many of these initiatives, other governments also became increasingly involved, particularly following the change in the US administration in 2008.

In addition to these national government-led initiatives, the second half of the 2000s also witnessed an increasing number of actions taken by subnational and non-state actors, often acting of their own volition. Subnational climate action in the United States entered the spotlight during the George W. Bush administration (2000–2008), with some states developing emissions trading systems by themselves, notably the Regional Greenhouse Gas Initiative and the Western Climate Initiative (Rabe, 2008). In addition, non-state actors started to play a role in a variety of areas, including carbon markets, company-to-company standard setting, greenhouse gas accounting and disclosure and project implementation.

From a scholarly perspective, the proliferation of new initiatives by a variety of public and private actors was described as a process of climate governance experimentation, with the term 'experiment' denoting any action that occurred outside the realm of multilateral treaties in the spirit of trial and error (Hoffmann, 2011: 17). The various cross-border collaborations between businesses, subnational authorities and civil society drew specific attention, and through databases and mapping efforts, a clearer picture started to emerge of the *transnational regime complex on climate change*' (Abbott, 2012, emphasis added; see also Bulkeley et al., 2012; 2014). The emerging governance system has been described as polycentric, with 'multiple governing authorities at different scales' exercising 'considerable independence to make norms and rules within a specific domain' (Ostrom, 2010: 552).

The mounting attention to action beneath and to the side of the international climate regime was not confined to academic circles: the role of non-state and subnational action has become part of the intergovernmental discussions under the UNFCCC since 2011, when parties launched a 'workstream' – i.e. a specific subset of negotiations – to discuss options for increasing ambition before 2020 in parallel to launching negotiations on a new climate agreement beyond 2020. Through technical papers and technical expert meetings, the prospective role of non-state and subnational action in scaling up ambition came in the spotlight. Attention for non-state climate action received a further boost at the COP in Lima in 2014, when the Peruvian government, together with the UNFCCC secretariat, launched the Non-State Actor Zone for Climate Action (NAZCA) platform, which registers commitments pledged by non-state actors worldwide. Moreover, together with the incoming French presidency of COP21 in Paris, an Action Agenda was announced to encourage and support new initiatives.

When the Paris Agreement was adopted in November 2015, it confirmed the direction already indicated by the Copenhagen Accord and Cancún Agreements, extending this system to the post–2020 period (Bodansky, 2016). Under the Agreement, parties are no longer subject to legally binding emission reductions. Instead, the system pins its hopes on a series of procedural obligations and the functioning of a mechanism to ratchet up ambition over time (Bodansky, 2016; Rajamani, 2016; van Asselt, 2016a). This mechanism would function as follows: (1) the direction of travel is determined by a long-term temperature goal to stay below 2°C and to pursue efforts to stay below 1.5°C, and a long-term goal of net zero carbon emissions by 2100; (2) parties submit new NDCs in five-year cycles; (3) new NDCs will have to go beyond previous ones and have to reflect a party's highest possible ambition; and (4) parties' efforts are subject to various types of (periodic) review, including a review of implementation through an 'enhanced transparency framework', a review of compliance through an implementation and compliance mechanism and a review of overall progress through a five-year global stocktake, starting in 2023. It is hoped that through an iterative process of submitting and reviewing NDCs, the long-term objectives will be achieved in due course.

Like the Copenhagen Accord and the Cancún Agreements before it, the Paris Agreement's architecture can be considered bottom-up in the sense that it leaves each country to determine the appropriate scale of action. However, just as the Kyoto Protocol was not fully top-down, the Paris Agreement is not fully bottom-up; importantly, by putting in place legally binding procedural obligations to submit NDCs, and subject their implementation to technical expert and peer review, the Agreement includes several top-down elements (van Asselt, 2016a).

Paris was also important because it strengthened the connections between multilateral governance on the one hand, and the growing number of experiments by non-state and subnational actors on the other. Although the Paris Agreement itself says remarkably little about non-state and subnational actors (Chan et al., 2016), the decision adopting the Agreement strengthens the engagement with non-state actors by specifying that technical examination processes (including one for adaptation as well as for mitigation) will continue up to 2020; calling for an annual 'high-level event' to take stock of non-state action and announce new initiatives; appointing two 'high-level champions' to ensure the successful execution of existing non-state actions and encourage new actions (UNFCCC, 2015).

After two and a half decades, it is becoming increasingly clear that achieving the lofty climate goals adopted by political leaders in 1992 cannot be the sole responsibility of the UNFCCC. Instead, we can observe an all-hands-on-deck approach: measures adopted under other regimes, such as the Montreal Protocol on ozone-depleting substances or the International Civil Aviation Organization (ICAO) on emissions from international aviation, are part of the international governance response (Keohane and Victor, 2011; Moncel and van Asselt, 2012; van Asselt and Bößner, 2016). Moreover, the variety of non-state and subnational action has become another key component of climate governance (Hale, 2016).

This raises the question of how such a multi-actored and multi-levelled governance system can be made to work. De Búrca, Keohane and Sabel (2013; 2014) have argued that 'global experimentalist governance' offers a promising model in this regard. The intellectual pedigree of the model rests in the debate on the EU – a block of countries that operates in a relatively cohesive way, despite a range of challenges such as diverging policy preferences, strong economic differences and a complex distribution of authorities. Proponents of the model have noted how, in certain, more complex policy domains, the EU has advanced successfully by offering a forum for coordination and the exchange of policy experiences. In so doing, lesson drawing from diversity occurs, and ideas about possible best practices emerge through explicit evaluation mechanisms. It has been argued that this model also has promise for climate governance. In the following, we test the extent to which the governance architecture established by the Paris Agreement resembles this ideal type of global governance, thus setting the stage for experimentation at national and local levels.

2.3 An Experimentalist Turn in Climate Governance?

Experimentalist governance has been defined as 'a set of practices involving open participation by a variety of entities (public or private), lack of formal hierarchy within governance arrangements, and extensive deliberation throughout the process of decision making and implementation' (De Búrca et al. 2013: 738). Sabel and Zeitlin (2008: 280) posit that the minimum scope conditions for experimentalist governance are a situation of 'strategic uncertainty' – i.e. a lack of knowledge about the best governance approach – and one of a 'polyarchic distribution of power' – i.e. no actor can impose its solution on others. In such situations, they argue, experimentalist governance offers a mode of governance that fosters learning. Although the concept of experimentalist governance originated and developed in contributions focused on the national or the EU level (e.g. Dorf and Sabel, 1998; Sabel and Zeitlin, 2008, 2010, 2012; Börzel, 2012), scholars have more recently raised the prospect of *global experimentalist govern-ance* (De Búrca et al., 2013, 2014).

According to its proponents, the ideal type of global experimentalist governance has four[1] interrelated elements. The first is the setting of open-ended framework goals, accompanied by metrics to measure progress, through a process that is open to a broad set of stakeholders, including civil society. Although the process should be participatory, the setting of such goals is done through a central authority. The presence of such a central authority underscores that experimentalist governance is not the same as purely bottom-up governance: experiments are – at least to some extent – managed and occur within a set of overarching rules, with a view to meeting the goals of the whole governance system. The second is decentralised implementation. Although moving in a broadly agreed direction, public and private actors at lower levels of governance may pursue their own goals in ways that they deem most fitting. It is their actions that are the 'experiments' in (global) experi-mentalist governance. According to experimentalist governance theory, a key driving force for actors to pursue the overall goals is the presence of a 'penalty default', which could consist of normative pressure through a loss of reputation or markets, sanctions imposed by law or the imposition of penalties by stronger actors on weaker actors (Sabel and Zeitlin, 2008; Sabel and Victor, 2017). A third element of global experimentalist governance consists of monitoring and reporting. The experimenting governance units – i.e. national and subnational governments, as well as non-state actors undertaking governing activities – are to monitor and report back on progress to the central authority, following common metrics. Furthermore, the units are expected to undergo a process of peer review in which their results are compared with those of others. If they do not make sufficient progress, 'the local units are expected to show that they are taking appropriate corrective measures, informed by the experience of their peers' (Sabel and Zeitlin, 2012: 179). Lastly, global experimentalist governance includes the evaluation of the goals set in the light of experience. This may lead to revising the goals and to launching a new cycle of experiments, monitoring, reporting and review.

Importantly, De Búrca et al. (2014: 478) add a normative dimension to their exploration of global experimentalist governance: if these elements are present, they posit, global experimentalist governance 'can constitute a form of governance that fosters a normatively desirable form of deliberative and participatory problem

---

[1] Following the earlier literature on experimentalist governance (e.g. Sabel and Zeitlin, 2010, 2012), we focus on four elements; De Búrca et al. (2013: 739) also mention 'openness to participation of relevant entities ('stakeholders') in a non-hierarchical process of decision making' as a separate element.

solving'. This suggests that it should be of interest to test the extent to which each of these elements features in the post–Paris governance architecture for climate change. This is what we set out to do in this section.[2]

2.3.1 Goals and Metrics

As a first step, experimentalist governance suggests that there needs to be 'thin consensus' – i.e. no fundamental disagreement on the existence of a problem and the need to address it (Sabel and Victor, 2017) – on the goals and associated metrics, which in turn requires a participatory decision-making process (De Búrca et al., 2013, 2014). To what extent is a thin consensus observable in the case of climate governance?

This question concerns the identification of the framework goals of climate governance. In a similar way that Sabel and Zeitlin (2008) identify 'good water status' and 'social inclusion' as framework goals that can be said to embody thin consensus in the context of various EU policy fields, the 'ultimate objective' of the UNFCCC to avoid 'dangerous anthropogenic interference with the climate system' (Article 2) could be said to constitute the goal of climate governance.[3] The goal was not just set on the basis of intergovernmental bargaining but also informed by scientific insights emerging in the late 1980s, including through the Intergovernmental Panel on Climate Change (IPCC) First Assessment Report from 1990 (Bodansky, 1993; Oppenheimer and Petsonk, 2005).

Over time, and drawing on emerging scientific insights on the causes and consequences of climate change, the UNFCCC's goal has been further specified in the form of long-term temperature goals. Initially, this was done through the gradual embrace of the 2°C goal (Jordan et al., 2013), which – before its inclusion in the 2009 Copenhagen Accord – had been adopted by the EU and the Group of Eight (G8) (Jaeger and Jaeger, 2011). More recently, the Paris Agreement offered more concrete guidance, by not only promoting the goal to stay well below 2°C but also suggesting to pursue efforts to stay below 1.5°C. In addition, the inclusion in the Paris Agreement of the goal to achieve net zero carbon emissions in the second half of this century offers further specificity.

---

[2] To be clear, this contribution is not the first to look at climate change in light of global experimentalist governance. However, Armeni (2015) focuses specifically on the governance of marine geoengineering – as opposed to climate governance more broadly – and Sabel and Victor (2016; 2017) and Keohane and Victor (2017) do not subject each of the elements analysed here to scrutiny.

[3] According to Keohane and Victor (2015: 207), the goal articulated in Article 2, as well as associated temperature goals, 'have been either too abstract or unachievable to specify near-term actions'. However, this sets a rather high, and hard to define, threshold for what constitutes experimentalist governance. It is unclear when a framework goal moves into the realm of the achievable, and other goals – e.g. social inclusion – are also inherently abstract and subject to interpretation.

The adoption of these goals in an international agreement can be cited as evidence for the presence of a thin (or even 'medium thick') consensus among nation states. However, it is notable that some of these goals, specifically the 2°C goal, are also embraced by a range of other actors. To give but a few examples: the science-based targets initiative – a collaboration between the UN Global Compact, the CDP (formerly the Carbon Disclosure Project), the World Resources Institute and the World Wide Fund for Nature – allows companies to showcase that they set targets in line with the 2°C goal;[4] several scenarios used to inform the International Energy Agency focus on staying below 2°C;[5] a 'climate test' has been developed by a set of environmental non-governmental organisations, which could help ensure that energy policy decisions are in line with the Paris Agreement's goals;[6] the need to keep temperature increases below 1.5°C became a rallying cry for civil society in the run-up to the Paris conference in 2015; and decarbonisation has been the underlying logic of new initiatives such as Track 0,[7] as well as social movements such as the fossil fuel divestment movement. In other words, decisions made by governments about long-term goals and processes to ensure their delivery have been influenced by many other actors, who in turn are affected by them.

The question of metrics is also related to the goals of governing. An initial response could be that the key metric for climate governance – at least in the mitigation domain – is a tonne of carbon dioxide equivalent (CO_2-eq.). Not only is the system established by the UNFCCC based on the accounting of greenhouse gas emissions in these terms but also this type of accounting resonates at other levels of governance, including for instance at the EU. The use of market-based instruments has led to the widespread use of this metric. However, this does not mean that there is consensus that this is the only – or the most appropriate – metric. The reductionist focus on CO_2-related metrics can be said to distract, or perhaps even undermine, the achievement of other (development) goals that are not captured by the metric (e.g. Frame, 2011). Moreover, for some actions, especially those undertaken by non-state actors, it will be hard to capture the results in terms of emission reductions. For instance, activities such as capacity-building, networking and sharing best practices and other types of information may not yield measurable result in terms of climate change mitigation, but may nonetheless be essential for effective climate action. As such, in the context of post–Paris climate governance, further metrics may need to be developed to capture the diversity of national and local actions and to foster comparability between them (Aldy and Pizer, 2016).

---

[4] http://sciencebasedtargets.org/
[5] www.iea.org/publications/scenariosandprojections/
[6] www.climatetest.org/
[7] http://track0.org/

2.3.2 Decentralised Actions

The next element of global experimentalist governance is decentralised implementation, offering individual, local or national governance units broad discretion. To what extent can climate governance be considered as decentralised?

This question ties in to an ongoing debate on whether the international climate regime can be considered 'top-down' or 'bottom-up' in nature. The Kyoto Protocol has been referred to as a quintessential example of a 'top-down' approach to international climate governance (e.g. Prins and Rayner, 2007; Rayner, 2010). However, Kyoto's targets and timetables were not imposed 'from above' to achieve the UNFCCC's objective, but rather were the subject of intense negotiations between countries; the Protocol's 'common objective' of 5.2 per cent greenhouse gas emission reductions between 1990 and 2008–2012 was simply the result of adding up what countries were willing to commit to. Likewise, the Copenhagen Accord – or the Paris Agreement for that matter – is not fully 'bottom-up'. While the Accord asked countries to make unilateral emission reduction pledges that were not the outcome of multilateral negotiations, the Cancún Agreements anchored the Copenhagen Accord's system of 'pledge and review' in the UNFCCC by elaborating the international reporting and review system developed under the Convention. In other words, the climate regime has always consisted of a hybrid of 'top-down' and 'bottom-up' elements (Dubash and Rajamani, 2010; Bodansky and Diringer, 2014), though elements of bottom-up climate governance have slowly moved to the foreground.

Importantly, none of the climate change treaties specifies the types of policies and measures countries should adopt to reach the general (e.g. staying below 2°C) or specific (e.g. a country's Kyoto target) goals. The Kyoto Protocol included concrete emission targets, but did not dictate how those targets needed to be met. It simply included a menu of optional policies and measures that countries could implement domestically to achieve their targets under the treaty. In other words, parties were offered significant discretion as to how (i.e. using which policies?) they would achieve their targets. The only limitation was that these goals were framed in the form of absolute, economy-wide emission targets. The Paris Agreement offers further flexibility and discretion by asking countries to pledge their goals on a five-yearly basis and indicate (roughly) what measures they seek to adopt in pursuit of those goals. Seen in this light, the NDCs – and the policies adopted to achieve them – constitute a variety of 'experiments' through which countries seek to meet the overall goals of the Paris Agreement.

However, as was outlined previously, it is now increasingly recognised that nation states are not (if they ever were) the only governance units responsible for the implementation of the goals. A wide variety of non-state actors – including civil society, business and research organisations and subnational authorities – have begun to take action on climate change. By the end of 2016, the UNFCCC's NAZCA platform listed 12,549 climate actions, including 2,138 companies, 238 civil society organisations and 2,578 cities in 118 countries (Yale Data-Driven Environmental Solutions Group, 2016). These different units also increasingly cooperate across borders. These transnational initiatives – such as the C40 cities network, the GHG Protocol, the Voluntary Carbon Standard, etc. – have also increased in number in recent years (Hoffmann, 2011; Bulkeley et al., 2012, 2014; Michaelowa and Michaelowa, 2017). Although states and international organisations such as the World Bank still sit in the driver's seat of many of these actions, or otherwise orchestrate them (Hale and Roger, 2014), these various initiatives significantly expand the universe of experiments in global climate governance.

As mentioned, the threat of 'penalty defaults' is seen as an important driving force for local authorities to engage in experimentalist governance. Can such a penalty default be deemed to be present in the context of global climate governance?

If a penalty default is narrowly conceived as sanctions in the case of non-compliance, the answer is, only to a limited extent. International climate agreements do not allow parties to resort to, for example, trade measures in case of non-compliance or non-participation (Sabel and Victor, 2017). Nonetheless, the Kyoto Protocol's compliance mechanism did allow for the imposition of some (limited) penalties, the strongest of which was the withdrawal of benefits in the form of excluding non-complying parties from participating in the Protocol's flexibility mechanisms (effectively raising the costs of compliance). However, as is documented elsewhere (Zahar, 2015), the compliance mechanism was unable to prevent blatant non-compliance of Canada in the late 2000s.

Penalty defaults may also materialise in other ways, including through pressure (naming and shaming) by environmental groups (Betsill et al., 2015). One way in which non-governmental organisations can increase pressure is by assessing the commitments made by states. Before the Paris conference, several of the then intended NDCs were heavily scrutinised by a range of research and civil society organisations. These analyses offered useful third-party perspectives on the information provided by governments, enabling stakeholders – including other parties, civil society organisations and concerned citizens – to form their own view on the level of ambition of the intended NDCs (van Asselt, 2016b). One example is the Climate Action Tracker, which releases regularly updated assessments of NDCs, labelling them from 'inadequate' to 'role model'.[8]

---

[8] http://climateactiontracker.org/

Whether penalty defaults exist in the context of non-state and subnational climate action is hard to verify, mainly because the motivations for such action are still unclear. These motivations can include many things: 'moral concerns, fear of new regulation (or the opportunity to secure first-mover advantages by shaping it), the pursuit of direct financial rewards, indirect or "non-climate" benefits (for example, reputational enhancement), and the satisfaction of consumer expectations' (Jordan et al., 2015: 979). In some cases – e.g. a government threatening to adopt new regulations in case voluntary action would not be forthcoming or sufficiently ambitious, as was the case in the EU for emissions from passenger cars – a penalty default could be deemed to be present for non-state action.

2.3.3 *Monitoring, Reporting and Review*

The third element of experimentalist governance is that individual governance units should monitor and report on progress and undergo a peer review (Sabel and Zeitlin, 2008). Is this the case in global climate governance?

Starting with states, this element of experimentalist governance requires a consideration of mechanisms for reporting and review under the international climate regime (Aldy, 2014; van Asselt, Sælen, and Pauw, 2015; Briner and Moarif, 2016; van Asselt et al., 2016; Gupta and van Asselt, forthcoming). Like most other multilateral environmental agreements, the UNFCCC, Kyoto Protocol and Paris Agreement contain provisions for reporting and review. The UNFCCC contains obligations for developed countries to report on emissions through greenhouse gas inventory reports and on their progress with implementation through national communications, both of which are subject to regular in-depth reviews by expert review teams. Developing countries also have to prepare these reports (albeit with more flexibility in terms of timing), but they are not subject to review.

New reporting and review processes were introduced by the Cancún Agreements in 2010. These required developed countries to submit biennial reports on their progress in achieving emission reductions and providing support to developing countries. These reports are subject to a process of 'international assessment and review' every two years. This includes a technical expert review and a 'multilateral assessment' – the latter involving a public question and answer session during UNFCCC meetings. The Cancún Agreements also specified that developing countries – except least developed countries and small island developing states – should submit new biennial update reports. These reports are subject to a process of 'international consultation and analysis', which resembles international assessment and review by including a technical expert part and a state-to-state 'facilitative sharing of view', but is to be non-confrontational and non-intrusive, and respectful of national sovereignty.

Building on the system of the Cancún Agreements (and eventually superseding them), the Paris Agreement establishes an 'enhanced transparency framework' applicable to all parties, but offering 'built-in flexibility' to account for parties' different capacities (Article 13). The framework will lead to a reporting and review system, through which parties have to report on their emissions and progress made towards their NDCs. Developed countries and other countries providing (financial, technology and capacity-building) support must further include information on the levels of support provided. Under the Paris Agreement, each party (except for least developed countries and small island developing states) needs to submit annual inventory reports, as well as biennial reports with information on the progress made towards NDCs. Like the arrangements established by the Cancún Agreements, the review process will include a technical expert review and a new process of 'facilitative multilateral consideration of progress'.

The foregoing discussion seems to suggest that the basic elements of monitoring, reporting and peer review, as required for global experimentalist governance, are in place – at least insofar as states are concerned. However, it is far less clear whether these elements operate effectively in practice. While the reporting record of developed countries is generally seen as adequate, albeit with some variation, the reporting record of developing countries is still troubling (Ellis and Moarif, 2015).

Moreover, although the multilateral assessment under international assessment and review and the facilitative sharing of views under international consultation and analysis – and in future the 'facilitative multilateral consideration of progress' – can be deemed peer review processes in which parties willingly undergo a review by other parties, their functioning in practice is still far from the ideal of experimentalist governance. The reviews are explicitly geared towards avoiding political judgements and rendering any proposed corrective measures impossible. Furthermore, the sensitivity of the process means that there is no opportunity for comparing the actions of some countries with those of their peers, or more generally assessing the effectiveness of parties' actions (Gupta and van Asselt, forthcoming).

In addition to covering the efforts by states, monitoring, reporting and review would also be needed for other governance units. This may include other international regimes (e.g. Stewart, Oppenheimer and Rudyk, 2013), some of which (e.g. ICAO) already regularly report to the UNFCCC. In addition, the NAZCA platform provides a system to keep track of non-state climate action. However, there is no obligation for most governance units to monitor or report – let alone review their actions. This can be explained by the need to ensure some distance between the central authority and the individual experiments, as underscored by the two high-level champions in their Global Climate Action Agenda Roadmap: '[I]t is important that [non-state climate actions] maintain some degree of independence from the UNFCCC process and governments in general.'[9] In other words, the suggestion is that a too stringent framework for monitoring, reporting and review may stifle experimentation, which probably is a concern harkening back to the notion that experiments can have multiple effects, not only effects in the realm of emission reductions (see also Betsill et al., 2015; Chan et al., 2015).

2.3.4 Evaluation

To come full circle, global experimentalist governance suggests a regular evaluation of experiences gained, leading to a possible revision of goals (and associated metrics).

The UNFCCC already contained a provision that mandated such an evaluation, instructing the COP to '[p]eriodically examine the obligations of the Parties and the institutional arrangements under the Convention, in the light of the objective of the Convention, the experience gained in its implementation and the evolution of scientific and technological knowledge' (Article 7.2[a]). This review led to the COP decision that launched the negotiations for the Kyoto Protocol, but was never followed by another review. The Kyoto Protocol also offered several opportunities for reviewing the adequacy of its commitments, but even though one of these processes led to the adoption of a second commitment period (2013–2020), new targets were based again on what developed countries were willing to do, rather than a review of those targets in light of experiences. Another review process was held between 2013 and 2015, with a mandate 'to periodically review the adequacy of the long-term global goal, ... in the light of the ultimate objective of the Convention, and overall progress towards achieving it' (UNFCCC, 2015: para. 138). This process – also known as the '2013–2015 review' – involved a 'structured expert dialogue', which brought together both parties and external experts.

Although the idea was that the 2013–2015 review would take place periodically, synchronised with the regular IPCC assessment reports, the adoption and rapid entry into force of the Paris Agreement meant that another periodic review, the global stocktake, will assume greater importance. The global stocktake is a process – the details of which remain to be determined – through which parties assess collective progress towards achieving the long-term goals of the Paris Agreement. The stocktake is to lead towards stronger ambition by parties as well as enhanced international cooperation. In this sense, the process is intended to lead to a revision of the goals adopted by parties for themselves, but not necessarily a revision of the overall goals of the Agreement itself. How exactly one can take

[9] http://newsroom.unfccc.int/climate-action/global-climate-action-agenda/#Roadmap.

stock of a very diverse set of actions – i.e. the NDCs – remains an important open question. Indeed, the same diversity encouraged by the Paris Agreement's emphasis on nationally determined approaches also means it will be challenging to 'add up' how parties are doing. This will be even more the case if the diversity of non-state actions are taken into account: even though initial efforts are made to evaluate such actions both ex ante (e.g. Hsu et al., 2016) and ex post (e.g. Chan et al., 2018), there are no clear benchmarks for assessing their performance (van der Ven, Bernstein and Hoffmann, 2017).

2.4 Conclusions: Prospects for Experimentalist Climate Governance

This chapter has examined the extent to which the global governance architecture post-Paris resembles the ideal type of global experimentalist governance that has recently been advanced by several scholars. That ideal type focuses on (1) the establishment of framework goals and associated metrics; (2) decentralised processes of implementation; (3) monitoring, reporting and review; and (4) evaluation and revision of the framework goals in the light of experience.

Starting with the first element, a thin consensus can be said to exist on broad framework goals in global climate governance. Whether the same can be said for metrics is far less clear, however, as the essential (but rather reductionist) focus on (tonnes of) CO_2 emissions reduced remains contested: it may simply not be appropriate to every one of the many climate actions noted previously.

Concerning the second element, decentralised implementation has always been part and parcel of the international climate regime, although the amount of discretion offered to states has greatly increased (note that contributions after Paris will be 'nationally determined'). Likewise, although not driven by a clear mandate by a central authority, a wide range of non-state and subnational climate actions means that climate governance has de facto become even more decentralised. Although some have suggested that a penalty default is required to drive climate action by states (Keohane and Victor, 2015; Sabel and Victor, 2017), it remains unclear to what extent it is really essential for national and subnational actions that are more polycentric in nature.

Moving on to the third element, monitoring, reporting and review has been a key feature of international climate cooperation for many years, although its effectiveness in terms of 'learning from difference' (cf. Sabel and Zeitlin, 2008) is, again, open to debate.

Finally, as regards the fourth element, the global stocktake offers an opportunity for parties to the Paris Agreement to reassess the appropriateness of their NDCs in light of emerging experiences of implementation. However, the diversity of national climate actions encouraged by the Paris Agreement will itself militate against quick stocktaking. In the absence of clear benchmarks of performance, one wonders how the plethora of non-state and subnational actions can ever be evaluated in a way that directly informs the revision of the longer-term goals, which thus far have been set according to scientific knowledge.

So, is climate governance starting to resemble the ideal type of experimentalist governance? Or, conversely, does global experimentalist governance actually represent a type of lowest-common-denominator governance, where the outcomes are largely dependent on what nation states and non-state actors are willing to put forward? Even if the answer to the first question is positive, we should nonetheless still critically assess the potential – and underlying premises – of that model. We conclude with four reflections in this regard.

First, the concept turns a blind eye towards the political dynamics underlying governance experiments, principally the role of power. Not all experiments are created equal, and some – particularly those driven by powerful state or non-state actors – may be purposefully created to influence the performance (either positively or negatively) of other experiments by challenging their goals or principles, or by promoting alternative means of implementation. An appreciation of the underlying politics of experimentation requires an in-depth analysis of the emergence, functioning and impacts of individual experiments, as will be offered in other chapters of this book. Doing so may shed further light on important questions, such as who participates in experiments, who leads them (see, e.g., Chan et al. [2018] for a discussion of North-South imbalances) and why they were first created.

Second, despite the seemingly bottom-up focus of experiments, the ontology of global experimentalist governance is in many ways still rooted in top-down thinking. Remember that the framework goals and metrics are to be set by a central authority (albeit with broad participation); the organisation of the overall evaluation is carried out at a central level; and the idea of a penalty default still shows that, at least to some extent, the governance system is still dependent on states (Jordan et al., 2015). Some have suggested that the emerging system is not just multi-levelled and multi-actored, but polycentric, with a great deal of action emerging spontaneously from the bottom up (Ostrom, 2010; Cole, 2015). However, that does not necessarily mean that polycentricity is a panacea: the evidence to back up the claim that such systems are necessarily more effective than globally oriented ones is still limited (Jordan et al., 2015).

Third, and perhaps somewhat ironically, for a governance system that has top-down elements embedded in it, there is potentially a lack of coordination inherent in experimentalist governance. Specifically, the management of experimentation does not extend to the design and conduct of controlled experiments that could eventually facilitate systematic learning. Although the absence of hierarchical steering is a purported feature of experimentalist governance, and excessive coordination may stifle experimentation (Betsill et al., 2015), the mere fact that (more) experiments are happening does not necessarily imply that common goals will be achieved or systematic lessons learned. This is particularly true in situations where there is a high degree of ambiguity about what the goals – and associated metrics – actually are. In our view, the performance of experiments is an empirical question, one which this book will shed further light on.

Fourth, we should be sensitive to the overall direction of global experimentalism, as well as issues of institutional design. As Hoffmann (2011: 25–26) has noted, the current wave of experiments are informed by a 'liberal environmental ethos', meaning that they stress the compatibility of economic growth and environmental protection and aim to enhance market forces. This, of course, is not the only ethos. If no one experiments with traditional, that is top-down, policy instruments such as regulation, our understanding of the alternatives will remain incomplete. And what of classical questions of institutional design? Who experiments? What information counts as relevant? Who decides whether the experiment was a success (McFadgen and Huitema, 2017)? In practice, not everyone can experiment, perhaps because of a lack of means, knowledge or political power. Therefore, how does the world build salient, credible and legitimate approaches (Cash et al., 2002) to climate governance? These and other important questions remain to be answered if scholars are to fully appreciate the nature and potential of experimentalist climate governance.

References

Abbott, K. W. (2012). The transnational regime complex for climate change. *Environment and Planning C: Government and Policy*, 30(4), 571–590.

Aldy, J. E. (2014). The crucial role of policy surveillance in international climate policy. *Climatic Change*, 126(3–4), 279–292.

Aldy, J., and Pizer, W. (2016). Alternative metrics for comparing domestic climate change mitigation efforts and emerging international climate policy architecture. *Review of Environmental Economics and Policy*, 10(1), 3–24.

Armeni, C. (2015). Global experimentalist governance, international law and climate change technologies. *International and Comparative Law Quarterly*, 64(4), 875–904.

Betsill, M., Dubash, N. K., Paterson, M., van Asselt, H., Vihma, A., and Winkler, H. (2015). Building productive links between the UNFCCC and the broader climate governance landscape. *Global Environmental Politics*, 15(2), 1–10.

Bodansky, D. (1993). The United Nations framework convention on climate change: A commentary. *Yale Journal of International Law*, 18(2), 451–558.

Bodansky, D. (2016). The Paris Climate Change Agreement: A new hope? *American Journal of International Law*, 110(2), 288–319.

Bodansky, D., and Diringer, E. (2014). *Building Flexibility and Ambition into a 2015 Climate Agreement*. Washington, DC: Center for Climate and Energy Solutions.

Börzel, T. A. (2012). Experimentalist governance in the EU: The emperor's new clothes? *Regulation & Governance*, 6(3), 378–384.

Briner G., and Moarif, S. (2016). *Enhancing Transparency of Climate Change Mitigation under the Paris Agreement: Lessons from Experience*. Paris: Organisation for Economic Co-operation and Development.

Brunnée, J. (2001). The Kyoto Protocol: Testing ground for compliance theories? *Heidelberg Journal of International Law*, 6(2), 255–280.

Bulkeley, H., Andonova, L., Bäckstrand, K., Betsill, M., Compagnon, D., Duffy, R., Kolk, A., Hoffmann, M., Levy, D., Newell, P., Milledge, T., Paterson, M., Pattberg, P., and VanDeveer, S. (2012). Governing climate change transnationally: Assessing the evidence from a database of sixty initiatives. *Environment and Planning C: Government and Policy*, 30(4), 591–612.

Bulkeley, H., Andonova, L., Betsill, M., Compagnon, D., Hale, T., Hoffmann, M., Newell, P., Paterson, M., Roger, C., and VanDeveer, S. (2014). *Transnational Climate Change Governance*. Cambridge, UK: Cambridge University Press.

Carrington, D. (18 January 2017). 2016 hottest year ever recorded – and scientists say human activity to blame. *The Guardian*.

Cash, D., Clark, W., Alcock, F., Dickson, N., Eckley, N., and Jäger, J. (2002). *Salience, Credibility, Legitimacy and Boundaries: Linking Research, Assessment and Decision Making*. Cambridge, MA: Kennedy School of Government, Harvard University.

Chan, S., Brandi, C., and Bauer, S. (2016). Aligning transnational climate action with international climate governance: The road from Paris. *Review of European, Comparative and International Environmental Law*, 25(2), 238–247.

Chan, S., Falkner, R., Goldberg, M., and van Asselt, H. (2018). Article title. *Journal title*, 18(1), 24–35. Effective and geographically balanced? An output-based assessment of non-state climate actions. *Climate Policy*.

Chan, S., van Asselt, H., Hale, T., Abbott, K. W., Beisheim, M., Hoffmann, M., Guy, B., Höhne, N., Hsu, A., Pattberg, P., Pauw, P., Ramstein, C., and Widerberg, O. (2015). Reinvigorating international climate policy: A comprehensive framework for effective nonstate action. *Global Policy*, 6(4), 466–473.

Cole, D. H. (2015). Advantages of a polycentric approach to climate change policy. *Nature Climate Change*, 5(2), 114–118.

de Búrca, G., Keohane, R. O., and Sabel, C. (2013). New modes of pluralist global governance. *New York University Journal of International Law & Politics*, 45(3), 723–786.

de Búrca, G., Keohane, R. O., and Sabel, C. (2014). Global experimentalist governance. *British Journal of Political Science*, 44(3), 477–486.

Dimitrov, R. S. (2010). Inside Copenhagen: The state of climate governance. *Global Environmental Politics*, 10(2), 18–24.

Doelle, M. (2016). The Paris Agreement: Historic breakthrough or high stakes experiment? *Climate Law*, 6(1–2), 1–20.

Dorf, M. C., and Sabel, C. (1998). A constitution of democratic experimentalism. *Columbia Law Review*, 98(2), 267–473.

Dubash, N. K., and Rajamani, L. (2010). Beyond Copenhagen: Next steps. *Climate Policy*, 10(6), 593–599.

Ellis, J., and Moarif, S. (2015). *Identifying and Addressing Gaps in the UNFCCC Reporting Framework*. Paris: Organisation for Economic Co-operation and Development.

Frame, D. J. (2011). The problems of markets: Science, norms and the commodification of carbon. *Geographical Journal*, 177(2), 138–148.

Gupta, A., and van Asselt, H. (forthcoming). Transparency in Multilateral Climate Negotiations: Furthering Accountability? *Regulation & Governance.*

Hale, T. (2016). All hands on deck: The Paris Agreement and non-state climate action. *Global Environmental Politics*, 16(2), 12–21.

Hale, T., and Roger, C. (2014). Orchestration and transnational climate governance. *Review of International Organizations*, 9(1), 59–82.

Hoffmann, M. J. (2011). *Climate Governance at the Crossroads: Experimenting with a Global Response after Kyoto.* New York: Oxford University Press.

Hsu, A., Cheng, Y., Weinfurter, A., Xu, K., and Yick, C. (2016). Track climate pledges of cities and companies. *Nature*, 532(7599), 303–306.

Hsu, A., Moffat, A. S., Weinfurter, A. J., and Schwartz, J. D. (2015). Towards a new climate diplomacy. *Nature Climate Change*, 5(6), 501–503.

Jaeger, C. C., and Jaeger, J. (2011). Three views of two degrees. *Regional Environmental Change*, 11(S1), 15–26.

Jordan, A. J., Huitema, D., Hildén, M., van Asselt, H., Rayner, T. J., Boasson, E. L., Forster, J., Schoenefeld, J., and Tosun, J. (2015). The emergence of polycentric climate governance and its future prospects. *Nature Climate Change*, 5(11), 977–982.

Jordan, A. J., Schroeder, H., Adger, N., Anderson, K., Bows, A., Le Quéré, C., Joshi, M., Mander, S., Vaughan, N., and Whitmarsh, L. (2013). Going beyond two degrees? The risks and opportunities of alternative options. *Climate Policy*, 13(6), 738–750.

Keohane, R. O., and Victor, D. G. (2011). The regime complex for climate change. *Perspectives on Politics*, 9(1), 7–23.

Keohane, R. O., and Victor, D. G. (2015). After the failure of top-down mandates: The role of experimental governance in climate change policy. In Barrett, S., Carraro, C., and de Melo, J. (eds.), *Towards a Workable and Effective Climate Regime.* London/Clermont Ferrand: CEPR Press and Ferdi, 201–212.

McFadgen, B., and Huitema, D. (2017). Are all experiments created equal? A framework for analysis of the learning potential of policy experiments in environmental governance. *Journal of Environmental Planning and Management*, 60(10), 1765–1784.

Michaelowa, K., and Michaelowa, A. (2017). Transnational climate governance initiatives: Designed for effective climate change mitigation? *International Interactions*, 43(1), 129–155.

Moncel, R., and van Asselt, H. (2012). All hands on deck! Mobilizing climate change action beyond the UNFCCC. *Review of European Community and International Environmental Law*, 21(3), 163–176.

Oppenheimer, M., and Petsonk, A. (2005). Article 2 of the UNFCCC: Historical origins, recent interpretations. *Climatic Change*, 73(3), 195–226.

Ostrom, E. (2010). Polycentric systems for coping with collective action and global environmental change. *Global Environmental Change*, 20(4), 550–557.

Prins, G., and Rayner, S. (2007). Time to ditch Kyoto. *Nature*, 449(7165), 973–975.

Rabe, B. G. (2008). States on steroids: The intergovernmental odyssey of American climate policy. *Review of Policy Research*, 25(2), 105–128.

Rajamani, L. (2013). Differentiation in the emerging climate regime. *Theoretical Inquiries in Law*, 14(1), 151–171.

Rajamani, L. (2016). Ambition and differentiation in the 2015 Paris Agreement: Interpretative possibilities and underlying politics. *International and Comparative Law Quarterly*, 65(2), 493–414.

Rayner, S. (2010). How to eat an elephant: A bottom-up approach to climate policy. *Climate Policy*, 10(6), 615–621.

Rogelj, J., den Elzen, M., Höhne, N., Fransen, T., Fekete, H., Winkler, H., Schaeffer, R., Sha, F., Riahi, K., and Meinshausen, M. (2016). Paris Agreement climate proposals need a boost to keep warming well below 2°C. *Nature*, 354(7609), 631–639.

Sabel, C. F., and Victor, D. G. (2016). *Making the Paris Process More Effective: A New Approach to Policy Coordination on Global Climate Change*. Muscatine, IA: Stanley Foundation.

Sabel, C. F., and Victor, D. G. (2017). Governing global problems under uncertainty: Making bottom-up climate policy work. *Climatic Change*, 144(1), 15–27.

Sabel, C. F., and Zeitlin, J. (2008). Learning from difference: The new architecture of experimentalist governance in the EU. *European Law Journal*, 14(3), 271–327.

Sabel, C. F., and Zeitlin, J. (eds.) (2010). *Experimentalist Governance in the European Union: Towards a New Architecture*. Oxford: Oxford University Press.

Sabel, C. F., and Zeitlin, J. (2012). Experimentalist governance. In Levi-Faur, D. (ed.), *The Oxford Handbook of Governance*. Oxford: Oxford University Press.

Stewart, R. B, Oppenheimer, M., and Rudyk, B. (2013). Building blocks for global climate protection. *Stanford Environmental Law Journal*, 32(2), 341–392.

UNEP (United Nations Environment Programme) (2016). *The Emissions Gap Report 2016*. Nairobi: United Nations Environment Programme.

UNFCCC (United Nations Framework Convention on Climate Change) (2010). *Decision 1/CP.16: The Cancun Agreements: Outcome of the Work of the Ad Hoc Working Group on Long-term Cooperative Action under the Convention*. Bonn: UNFCCC.

UNFCCC (2015). *Decision 1/CP.21: Adoption of the Paris Agreement*. Bonn: UNFCCC.

van Asselt, H. (2014). *The Fragmentation of Global Climate Governance: Consequences and Management of Regime Interactions*. Cheltenham, UK: Edward Elgar.

van Asselt, H. (2016a). International climate change law in a bottom-up world. *Questions of International Law*, 6, 5–15.

van Asselt, H. (2016b). The role of non-state actors in reviewing ambition, implementation, and compliance under the Paris Agreement. *Climate Law*, 6(1), 91–108.

van Asselt, H., and Bößner, S. (2016). The shape of things to come: Global climate governance after Paris. *Carbon & Climate Law Review*, 10(1), 46–61.

van Asselt, H., Sælen, H., and Pauw, P. (2015). *Assessment and Review under a 2015 Climate Change Agreement*. Copenhagen: Nordic Council of Ministers.

van Asselt, H., Weikmans, R., Roberts, T., and Abeysinghe, A. (2016). *Transparency of Action and Support under the Paris Agreement*. Oxford: European Capacity-Building Initiative.

van der Ven H., Bernstein, S., and Hoffmann, M. (2017). Valuing the contributions of nonstate and subnational actors to climate governance. *Global Environmental Politics*, 17(1), 1–20.

Yale Data-Driven Environmental Solutions Group (2016). *Taking Stock of Global Climate Action*. http://datadriven.yale.edu/wp-content/uploads/2016/12/Data_Driven_Yale_Taking-Stock-of-Global-Climate-Action_Nov_2016_final.pdf [accessed 4 October 2017].

Zahar, A. (2015). *International Climate Change Law and State Compliance*. London: Routledge.

Part I

Experiments

Exploring Innovations in Climate Governance

3

Anchoring and Mobility of Local Energy Concepts

The Case of Community Choice Aggregation (CCA)

LUÍS CARVALHO AND IRINA LAZZERINI

3.1 Introduction

Greening the power supply is one of the most pressing climate challenges world-wide. Several try-outs and the occasional take-off of renewable power sources have occurred over the last decade, nudged by multiple actors and facilitated by several policies and governance arrangements. Recent work in the field of socio-technical transitions has started to pay attention to the nuanced geographies involved on these and other types of climate and energy-related experimentation (Coenen, Benneworth and Truffer, 2012; Raven, Schot and Berkhout, 2012), here understood as deliberate interventions to seed change in established socio-technical systems by testing out alternative (energy) concepts, processes and practices (e.g. Bulkeley and Castán Broto, 2013; Turnheim, Kivimaa and Berkhout, Chapter 1).

Such a spatially informed view has already brought new insights on how green power alternatives unfold, gain traction and eventually bring change to incumbent socio-technical regimes (Hansen and Coenen, 2015). Those include new conceptual entry points to understand how knowledge is produced and travels across multiple geographies, as well as on how the material, social and political texture of places influence those processes – for example, by accelerating or creating friction within (Truffer, Murphy and Raven, 2015).

This chapter contributes to this recent stream and discusses the relevance of such a geographical lens to understand how new energy concepts and climate governance innovations gain traction and actually move beyond single experiments. It builds on previous research that combines transition studies and economic geography to propose that a useful framework to better understand how experiments become more embedded involves looking at their journeys as 'spatialized' sequences of anchoring, recombination and mobility (Binz, Truffer and Coenen, 2016; Carvalho, Mingardo and Van Haaren, 2012; Carvalho, Santos and Van Winden, 2014b; Sengers and Raven, 2015). Such an analytical perspective allows for a more explicit grasp on which concrete places (e.g. towns, cities, regions) matter in accelerating the diffusion of new climate alternatives, as well as how (and by whom) experiments and new practices are transported and transformed beyond their initial sites and configurations.

This chapter illustrates such a spatialized view with the case of Community Choice Aggregation (CCA) in the United States. CCA is a grassroots and community energy innovation (e.g. Seyfang et al., 2014) through which local governments aggregate demand and procure electric power on behalf of its constituents (California Energy Commission, 2006; Faulkner, 2010). CCA was first experimented in a few small towns in Massachusetts and Ohio in the late 1990s and focused on cost savings. During the 2000s, the concept was 'transported' to California where it blended its original principles of carbon trading with green activism and local empowerment, leading to the formation of municipal-driven 'buyer's clubs' of green power. After a sluggish start in which many hurdles had to be overcome, two counties in the San Francisco Bay Area (Marin and Sonoma) and the city of Lancaster (Los Angeles County) have, at the time of this writing, fully operating CCA programmes offering 100 per cent green energy options, while many other places in California are in the process of implementing similar schemes. Recently, the concept has travelled back eastwards and is being implemented in other municipalities in the states of Illinois, New Jersey and New York.

This chapter traces back to the origins of the CCA concept and zooms into the ways it has been experimented with, anchored and recombined in Marin County before gaining traction and moving to other places. It explores the concrete social, political, cultural and ecological features of Marin that facilitated the legitimation and anchoring of the CCA innovation in this jurisdiction; moreover, it compares the experience in Marin with the situation in the city of San Francisco, in which local politics and heterogeneous social preferences had been holding down the adoption of a long-planned CCA. Subsequently, this chapter describes the ways through which CCA concepts became recombined in place (i.e. leading to new visions, expectations and lessons from experience) and the actions of several actors – consultants, lawyers, energy companies, climate protection advocates, etc. – moving it to other places in California and beyond. The underlying evidence in this chapter draws on twenty-two in-depth interviews conducted during March 2014 for a study on energy transitions in San Francisco's Bay Area (Carvalho, Lazzerini and Tuijl, 2014a) and on three additional follow-up interviews conducted during February–March 2016, complemented by extensive desk research.

Section 3.2 of this chapter reviews recent research on the geography of sustainability transitions, in particular of niche development processes, and brings forwards the notions of anchoring and mobility as a complementary framework to look at niche development from a spatially nuanced perspective. Section 3.3 illustrates such a framework with the case of the embedding of CCA in the United States. Section 3.4 concludes by recapping implications for socio-technical transitions theory and the embedding of experiments and climate governance innovations.

3.2 Conceptual Background

3.2.1 Niche Development and the Geography of Sustainability Transitions

The literature that studies long-term change in socio-technical systems, such as energy systems, argues that broader change is preceded by a number of experiments in which new socio-technical configurations are incrementally tried out (e.g. Geels, 2002; Kemp, Schot and Hoogma, 1998). Experimentation often benefits from some degree of protection – for example, through the strategic action of committed actors, institutional and policy arrangements – by which new proto-regimes and socio-technical configurations are tested and gain leeway (e.g. Hoogma et al., 2002). This is so because alternatives are often incipient and immature, requiring learning (on alternative technologies; regulatory and legal procedures; user behaviours; policy processes; economic, societal and environmental effects, etc.), as well as new networks of advocates and the articulation of expectations and visions (Schot and Geels, 2008).

Despite the pivotal role attributed to experimentation, the literature on niche development argues that prevailing socio-technical configurations can hardly be challenged by single experiments in particular sites and locales (e.g. Kemp et al., 1998; Smith and Raven, 2012). Geels and Raven (2006) described how sequences of concrete and localized experiments can interact with one another to give rise to a more 'influent' global niche level, understood as an emerging field or a community-level structure of shared rules – such as common narratives, problem agendas, expectations, search heuristics, strategies, etc. According to this perspective, global niches coalesce around learning trajectories with heightened capacity to infuse change in dominant socio-technical regimes. Geels and Deuten (2006) conceptualized global niche formation as a process evolving from fragmented local and inter-local experiments in context towards trans-local and global aggregation stages, in which a niche's knowledge and shared rules become increasingly mobile and 'cosmopolitan' and, thus, non-spatially situated. As Schot and Geels (2008: 543) put it:

[Niche] developments may start with one or a few projects, carried out by local networks of actors, who are interested in innovations for idiosyncratic or local reasons. The cognitive rules ... that guide these projects are initially diffuse, broad and unstable. Local projects form test beds for these diffuse ideas. ... If learning processes in local projects are compared and aggregated, the cognitive rules at the more global niche level may gradually become more articulated, specific and stable.

This conceptualization provides valuable insights into how the socio-cognitive orientation of the actors involved in socio-technical experimentation are formed and retained, driving (or hampering) experimentation further by shaping search directions, rhetoric, content, policy designs, expectations, etc., in an iterative fashion (Geels and Deuten, 2006). Moreover, it calls attention to the role of intermediary actors' activities and de-localizing parts of the knowledge derived from localized experiments (see also Seyfang et al., 2014, for the case of grassroots energy innovations). Such intermediaries can operate as part of official governance structures (Kivimaa, 2014) or as non-state private governance (e.g. Bessant and Rush, 1995). Yet, despite its advantages, the local-global niche model has been criticized for saying little about where concretely new socio-cognitive frames become formed, through which channels they travel to become 'global' and how they become re-embedded across places (Hansen and Coenen, 2015).

This sympathetic criticism has been part of a more general effort to bring geographical insights into the study of sustainability transitions (e.g. Coenen et al., 2012; Raven et al., 2012; Truffer and Coenen, 2012). Geographically informed studies on sustainability transitions start from the recognition that many transition processes – namely the early formation and consolidation of niche experiments – often bypass the national level and other discrete spatial units and progress through nuanced geographical configurations to gain traction, involving, among others, concrete cities, regions, transnational relations and other non-contiguous geographies revealed by the networks and agency of mobile actors (e.g. Berkhout et al., 2010; Binz, Truffer and Coenen, 2014; Carvalho et al., 2012; Wieczorek, Raven and Berkhout, 2015). By bringing on board key dimensions of geographical studies – such as place, space and the relations between agents and their environments – these studies argue that taking geography seriously can contribute to a better understanding about how niches and experiments form and unfold, providing stronger analytical insights about strategies to influence their development further.

Among others, two dimensions have deserved particular attention in this literature, namely the notions of *socio-spatial embedding* and *multi-scalarity* (Coenen et al., 2012; Truffer et al., 2015). First, notions of socio-spatial embedding contribute to understanding why certain experiments, niche activities and ultimately transition processes are more successful in some places than in others. This calls for an understanding of places not simply as 'sites' where experiments unfold but as distinct assemblages of actors, networks and institutions with cultural, social and political texture, influencing the direction of experimentation and its ability to gain traction. Second, engaging with notions of multi-scalarity can contribute to highlighting the actors, processes and geographical scales through which experiments become intertwined and connect to one another towards the formation of broader alternatives to dominant socio-technical regimes. Overall, by bringing in explicit geographical notions, these studies suggest that sustainability transitions evolve and gain traction through intrinsically spatial processes, and, therefore, the concrete places where experiments form – and travel to – may largely influence their journeys towards becoming more than experiments.

A recent study by Sengers and Raven (2015) made use of several insights from economic geography – for example, on the spatial organization of production networks, the geography of innovation and urban policy mobilities – to 'spatialize' the local-global niche development model, contributing to explain how new socio-technical configurations gain a foothold and move beyond single experiments. Through a study of bus rapid transit systems worldwide, the authors concluded that experiments are embedded not only in a-spatial community fields but also in concrete territorial settings (e.g. in cities), whose material, relational, cultural, political and symbolic features can contribute to nurturing experiments and new propositions (but also potentially decreasing expectations about the advantages of experimentation). Moreover, they stress that insight and learning about experiments and new socio-technical configurations do not just travel and aggregate automatically or without friction, but through deliberate actions of concrete actors that bridge between local contexts and selective global networks (of corporate interests, policy advice, consultancy and technocracy, not-for-profit advocacy, etc.). As recognized by the Sengers and Raven (2015: 170),

the exchange of codified knowledge, lessons and hands-on expertise requires that the actors put in dedicated efforts to build and maintain communicative channels in order to receive new knowledge and transmit to the wider world the vibrancy of a particular place and the niche experiments that shape and are shaped by it.

The 'spatialization' of the local-global niche model highlights the distinct nature of local contexts and the selectiveness of the networks mobilized by specific actors – which are simultaneously mobile and linked to concrete places – moving concepts, lessons and other shared rules in space. Hence, concrete cities and regions may be not just sites of diffusion for new socio-technical configurations but also social and politically rich arenas of mediation where an experiment's content and visions are negotiated, reconfigured and recombined.

3.2.2 Moving Beyond Experiments through Sequences of Anchoring and Mobility

Building on the previous work to spatialize niche development processes, it can be argued that the road for climate-related experiments to become more than experiments involves spatial circuits through which learning, new networks and associated expectations are formed, transformed and (re-)assembled in concrete places before becoming mobile again.

Such a framework can be grounded on the notions of anchoring and mobility. Like other concepts from economic geography recently applied in the study of sustainability transitions (e.g. Coenen et al., 2010; Sengers and Raven, 2015), the notion of anchoring has been used to understand change in territorial production and innovation systems under heightened mobility of production factors, notably knowledge (Crevoisier and Jeannerat, 2009; Vale and Carvalho, 2013). This work posits that under such a circulatory paradigm, the development of localized innovation systems became increasingly reliant on the capacity of a place to engage with resources and knowledge developed elsewhere, being that different localities will mobilize and integrate mobile knowledge in different ways (De Propris and Crevoisier, 2011).

According to De Propris and Crevoisier (2011: 173), anchoring is the process through which different places manage to 'access, interact and capture knowledge, information, ideas or any form of tangible or intangible asset from other places', relying on the (place-based) capacities of their actors, networks and institutional settings. Anchoring makes sense only in relation to its twin notion of mobility – that is, of movement of tangible and intangible assets across space. Under this perspective, knowledge is still produced and nurtured in concrete socio-spatial contexts, and anchoring is the capacity of other places to re-contextualize/ assimilate complex knowledge pieces and diffuse them locally, before they eventually become mobile again and flow to other contexts. Notably, and notwithstanding the ability of some forms of knowledge to de-multiply and travel across space with little friction – as for the global niche 'cosmopolitan' knowledge (Geels and Deuten (2006) – recent transition studies on community energy initiatives (Seyfang et al., 2014) also stress that some territorially produced forms of tacit knowledge and trust are not easily moved to new settings. This heightens the relevance of anchoring in shaping and mediating the absorption of knowledge flows across spatial settings.

The concept of anchoring resembles other notions of embeddedness (Turnheim et al., Chapter 1). Yet, while embeddedness broadly refers to stable relationships between agents/actions and structures (e.g. social, institutional and cognitive environments) 'belonging' to concrete places of origin, 'anchoring embodies the idea that there exists an element of mobility between places, namely a tendency to open or a movement towards a new context' (De Propris and Crevoisier, 2011: 174), which implies a dynamic and frequent re-contextualization and transformation of knowledge and other mobile resources in the process. Notably, the capacity to re-contextualize mobile resources depends on the actors and relational assets in place, as well as on deliberate actions and policy efforts to strike the necessary alignments between localized assets and mobile resources (e.g. Coenen, Raven and Verbong, 2010; Vale and Carvalho, 2013). These abilities become increasingly pivotal for climate experimentation and sustainability transitions, particularly as a growing number of experiments and niche development processes become linked to trans-local and transnational connections in many ways (e.g. Carvalho, 2015; Hansen and Coenen, 2015, Wieczorek et al., 2015).[1]

The aforementioned, geographically informed notions of anchoring and mobility have been recently applied in studies at the intersection of economic geography and sustainability transitions. For example, Binz et al. (2016) analysed how Beijing became a relevant node in on-site water recycling solutions by anchoring external knowledge, namely by actively creating new local markets, mobilizing investment, fostering local spin-offs and locally legitimizing the solution (e.g. in local hotel segments). Yet, to put those processes in motion and circulate newly created knowledge, the role of foreign international organizations, global companies and returnees was fundamental. In a similar vein, Carvalho et al. (2012) analysed how three cities (Curitiba, Hamburg and Goteborg) developed distinct assets that made them relevant international hubs for the creation, recombination, anchoring and diffusion of knowledge and ideas of sustainable mobility over time. Likewise, the fluid and rich mobility of new concepts and experiments to other places relied on the actions of globally connected as well as locally embedded actors, such as transnational corporations (car makers and energy utilities) and well-connected mayors. Moreover, in a different vein, the study by Sengers and Raven (2015) showed that a concrete failure to adopt a rapid bus system in Bangkok became a burden to the actors aggregating knowledge on it, hampering the diffusion of the concept in other places.

The previous studies provide insights into how sustainability and climate-related experiments may gain traction (or stall), namely by highlighting the role of spatialized sequences of (place-based) anchoring and mobility. This is because anchoring does not usually mean pure diffusion but the negotiation and re-contextualization of knowledge, ideas and practices, influenced by material, cultural, symbolic, social and political contexts of the places where new concepts move to and flow from. Such a perspective calls attention to concrete places that matter in accelerating the diffusion of new climate alternatives and also to the actors – territorially embedded and spatially mobile – by whom experiments and new practices are transported and transformed beyond their original sites and configurations.

---

[1] Other related notions of anchoring and mobility – both as metaphors and detailed constructs – have been recently put forward in transition studies (e.g. van Buuren et al., Chapter 8; Elzen, van Mierlo and Leeuwis, 2012). In these formulations, 'anchoring' has been used to study emerging processes of linking and forging (initially fragile and evolving) connections between experimentation and regime structures. Yet, in these approaches, experiments, niches and regimes are conceptualized as geographically neutral notions, and processes of anchoring are understood as more temporal than spatial.

The next section illustrates such a framework of spatialized sequences of anchoring and mobility with the case of CCA in the United States.

3.3 Spatial Circuits of a Local Energy Concept: Community Choice Aggregation (CCA)

3.3.1 State of Play

CCA is a voluntary yet law-regulated arrangement that allows local jurisdictions in the United States (e.g. municipalities, counties) to aggregate the electricity demand of its constituents and directly procure electricity to meet those demands (California Energy Commission, 2006; Faulkner, 2010).[2] Local authorities may deploy CCA with many objectives in mind, but typically to reduce electricity rates and/or to diversify energy sources towards more climate-friendly portfolios. Many states increasingly see CCA, together with other voluntary green power initiatives, as a pivotal instrument to comply with their legal renewable power standards (NREL, 2015). CCAs are often described as 'hybrids' between investor-owned and municipal utilities. Under CCA, electricity is usually procured and purchased by a public agency (representing the local jurisdiction), but the transmission, distribution and customer service is provided by the incumbent utility. As articulated by Asmus (2009: 311), such a system

> provides an easier way to change the content of the power supply [in cities or regions] without taking on the burden of managing the power lines, collecting bills and the divisive politics involved in the typically highly contested (and expensive) municipalization process.

CCA tends to face opposition from incumbent utilities. It implies change in business models and governmental energy policy as well as users' cultural frames on electricity generation and provision. Moreover, municipalities developing CCA may face financial (long-term contracts) and regulatory (state-level policy) risks, being that CCA formation can easily be blocked by local politics, often requiring majority votes. For these reasons, CCA requires considerable experimentation to challenge entrenched energy systems. New knowledge on legal affairs, political processes, energy procurement, business models, etc., as well as broad networks of advocates, new narratives and expectations about its societal advantages, has to be developed.

Despite these difficulties, CCA is currently a well-established alternative in the electricity system of seven states (Figure 3.1). It is estimated that about 5 per cent

---

[2] For a more detailed analysis of the context, evolution and structuration processes of CCA in California and Marin County, please see Carvalho et al. (2014a).

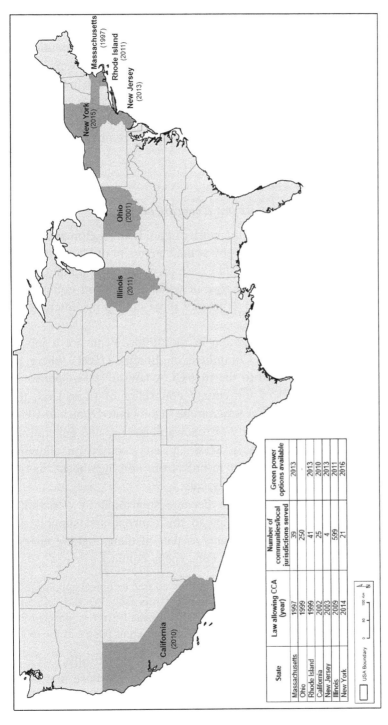

State	Law allowing CCA (year)	Number of communities/local jurisdictions served	Green power options available
Massachusetts	1997	39	2013
Ohio	1999	250	-
Rhode Island	1999	41	2013
California	2002	25	2010
New Jersey	2003	4	2013
Illinois	2009	599	2011
New York	2014	21	2016

Figure 3.1. States in the United States with active CCA programmes, March 2016 (launch year of the first CCA programme in brackets).

Source: Authors' elaboration, based on data from LEAN Energy.

Note: Data for Ohio, Rhode Island and New Jersey updated October 2013. For Massachusetts (updated October 2014), California and Illinois (February 2016) and New York (December 2015) (complementary data sources were used: NCSL, 2015; U.S. Department of Energy (2016); Centre for Climate Protection, 2016). In late 2015, Pennsylvania, Minnesota and Utah were in the process of developing CCA-supportive legislation (NREL, 2015).

of the U.S. population was served under CCA initiatives in 2016, covering about 1,000 local jurisdictions (Local Power, 2016; Figure 3.1). The first CCA initiatives date from the late 1990s and were implemented in small municipalities in Massachusetts and Ohio. The CCA concept then travelled to California in the early 2000s; Marin County launched the first operative CCA initiative in California in 2010, with other local jurisdictions following suit (Figure 3.2). Since 2010, CCA initiatives have been gaining traction in California and beyond. Moreover, their visions have been evolving from pure cost savings towards greening power supply and the localization of electricity production.

3.3.2 *Origins and First Experiments*

The origins of CCA can be traced back to the work of Paul Fenn, a climate activist and graduate student on history and philosophy at the University of Chicago in the early 1990s. Triggered by the work of Ronald Coase on commodity trading, Fenn thought about an alternative model for energy in which liberalized commodity trading would be mediated by local governments representing the interests of smaller communities (Sane Society, 2012).

In the midst of rising energy liberalization in the 1990s, Fenn got a job as Director of the Massachusetts Senate Committee on Energy (1994) where he drafted the senate bill that would lead to the first CCA law in 1997, sponsored by a senator and with the support of the county counsellor of Cape Cod, the jurisdiction that would form the first CCA programme in the United States in 1997. Yet, at this time, and due to the immaturity of renewable energy and generalized lack of interest at the state level, CCA in Massachusetts focused on allowing groups of municipalities to purchase power in bulk from non-incumbent energy providers to achieve rate savings for its constituents.

Following this isolated experiment, Fenn became sponsored by a national non-governmental organization (NGO) to spread the concept and support the development of CCA legislation in other states as part of their ongoing energy restructuring processes. As explained by Paul Fenn (pers. comm.),

We worked [in other States] through local NGOs and provided free technical support [e.g. on regulatory proceedings, new business models, political process]. We did selection matrixes to identify States with a combination of high [electricity] rates, progressive legislators and places where we could get local support to sit different people at the table.

Hence, after Massachusetts, Fenn became involved in drafting senate bills for CCA legislation in Ohio (approved in 1999) and New Jersey (2003), whose first CCA programmes would also focus on costs savings. In the early 2000s, Fenn moved to California (where he was originally from) and introduced the CCA concept by

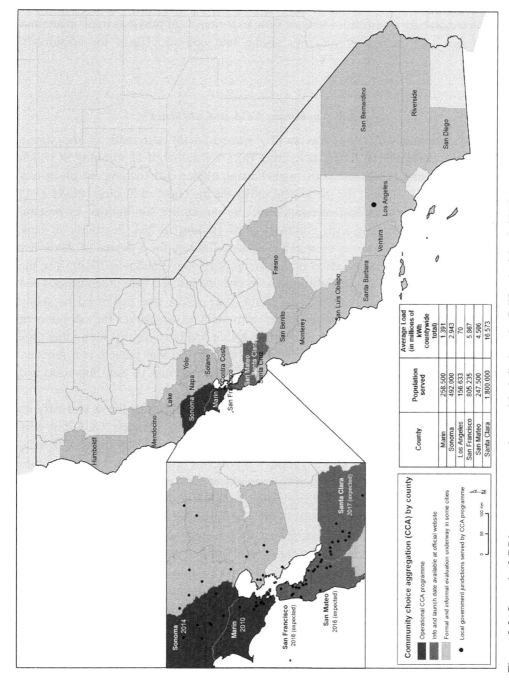

Community choice aggregation (CCA) by county		
■	Operational CCA programme	
▨	Info and launch date available at official website	
▨	Formal and informal evaluation underway in some cities	
•	Local government jurisdictions served by CCA programme	

County	Population served	Average Load (in millions of kWh countywide total)
Marin	258.500	1.391
Sonoma	492.000	2.943
Los Angeles	156.633	70
San Francisco	805.235	5.867
San Mateo	247.500	4.506
Santa Clara	1.800.000	16.573

Figure 3.2. Spread of CCA programmes and served jurisdictions in California, March 2016. *Source:* Authors' elaboration, based on data from the Centre for Climate Protection (2016).

authoring the state's CCA law in 2002 (Assembly Bill 117), with the sponsorship of a state assemblywoman (Local Power, 2016). By that time, the new CCA law reflected already a number of lessons from experiences of previous implementation processes (e.g. regarding majority voting and opt-out clauses for costumer's shifts).

3.3.3 Anchoring CCA in California

The Californian CCA law was swiftly passed in the aftermath of the state´s electricity deregulation fiasco, benefiting from a window of opportunity in which severe droughts combined with market manipulation led to rampant electricity prices and the bailout of the incumbent utility Pacific Gas and Electric (PG&E). As Californian regulators had re-established the possibility for utilities to procure electricity through long-term contracts (thus leading to less flexibility in consumer's choice), CCA was seen as a way to still allow for a degree of flexibility and competition in the selection of electricity providers (Faulkner, 2010). Yet, the expectations about CCA in California soon became infused with green visions and linked to other climate experiments in the state. CCA was perceived by many politicians and climate activists as an instrument to accelerate the adoption of renewable energy portfolios in California, which had been legally set by that time at 20 per cent for 2017.[3] Soon after Assembly Bill 117 was passed, two jurisdictions filled resolutions to experiment with CCA – the city of San Francisco and Marin County (see Figure 3.2).

Yet, CCA's anchoring processes would take rather different forms and varying degrees of success in the two jurisdictions. In Marin, the new CCA law was brought to the attention of county representatives by Sustainable Fairfax (a local grassroots organisation). This triggered a number of task forces feasibility studies and political and public discussions to implement CCA across Marin's municipalities. The aim was not to decrease cost (though it also occurred) but to 'move faster' than PG&E in greening electricity portfolios. To do that, and after multiple political struggles, popular votes, pros and cons campaigns and negotiations with PG&E and the Californian energy regulator (Carvalho et al., 2014a), Marin County started to directly procure electricity in the market through competitive tendering processes in 2009, as well as through buying green energy credits; this was done with the support of a newly formed local public agency – Marin Clean Energy.[4] In order to secure lower prices and increase green content, Marin signed contracts

---

[3] In 2016, this target was legally set at 40% for 2024 as part of the so-called State's Renewable Portfolio Standard (RPS).
[4] During the process of championing and lobbying for CCA in California, a state law was passed that separated the financial risk of the newly formed agency from the municipalities that formed it.

with large generation companies located in California and adjacent states, such as Shell North America. Marin Clean Energy started by offering two electricity packages: 50 per cent green and 100 per cent green, the latter with a slight price increase vis-à-vis PG&E.[5] During 2011 and 2014, other municipalities in Marin and in contiguous counties (Figure 3.2) joined Marin Clean Energy. At the same time, with steady implementation, Marin progressively expanded its CCA vision and concept. By reinvesting the profits of CCA electricity sales locally, Marin Clean Energy started to move from procuring wholesale green energy in the market to buying credits towards incentivizing energy efficiency and the localization of energy production within the county's boundaries (e.g. distributed solar production).

Apart from local political leadership and the presence of closely knit networks of politicians and green activists, the socio-spatial features of Marin County contributed to explain the new directions followed by the CCA concept, as well as why Marin succeeded to move it beyond the purely experimental stage while other early attempts in California failed. First, the physical context of Marin (mostly small municipalities with stable residential energy markets) provided enough scale economies and reduced financial risks by making energy loads predictable and manageable. Second, the socio-economic bracket and political preferences in Marin were considerably homogeneous ('wealthy, liberal and green'), making it easier to aggregate constituents' choices and decrease sensitivity to the uncertainty surrounding CCA. Third, Marin had a strong tradition of grass-roots movements and rallying environmental groups favouring renewable energy, yet they were highly sensitive to landscape preservation. In an interview, a representative of Marin County explains:

One way to become [even] greener was through renewable energy procurement; ... policies to reduce car use would be politically unfeasible due to the hilly geography and the fact that most families own more than one car. [And contrarily to Marin], a green procurement contract with Shell would never pass in other cities like Berkeley, who are strong opponents of oil corporations in any form. In Marin, landscape preservation [e.g. from wind turbines] was considered more important.

The city of San Francisco filled in a resolution to explore the formation of CCA roughly at the same time as Marin County did (2004), following similar visions – i.e. using CCA to increase green energy in the mix – initially with the in-house collaboration of Paul Fenn. To this effect, San Francisco Public Utilities Commission set up CleanPowerSF, with the support of the city's board of supervisors. Yet, CleanPowerSF proposed using CCA to provide 100 per cent renewable energy-

---

[5] It is estimated that PG&E's green content in their portfolio (including wind, solar, biomass, biogas and hydro) was about 33% in 2015 (fieldwork).

based electricity, fully generated in California, which was estimated to drive the prices up by 10–15 per cent (with expected opt-out rates of 70 per cent). As expressed by an interviewee, this would be like 'sprinting the whole marathon'. Moreover, and because of this, San Francisco's CCA plans lacked early societal and political support and soon became mired in politics (Lagos and Baker, 2013). A (rate-reducing) contract with Shell was seen as too controversial; moreover, San Francisco-based PG&E labour unions lobbied locally against the plan, fearing job losses. All this led to political infighting within the city, with the mayor and the Public Utilities Commission opposing the plan and the board of supervisors in favour of the initiative. In order to overcome these barriers, CleanPowerSF started loosening their targets on green content, while insisting on using CCA as a tool to localize energy generation and create green jobs, namely through active photovoltaic and energy efficiency policies (in line with Marin's recent developments).[6]

Over the course of about ten years, CCA policy criteria has been anchored and re-contextualized in California towards a whole new concept, called the California model or CCA 2.0 (Local Power, 2016) – i.e. departing from rate savings to include renewable power procurement and, increasingly, the localization of electricity generation, energy efficiency and job creation in the community. As seen, this was facilitated by an overall fertile ground to breed these ideas in California, as well as by the particular political, relational, material and cultural features of the places in which it became anchored earlier (namely in Marin). During this course of events, CCA became more than an experiment and deeply permeated the energy system in California. In the process, broad learning about CCA took place (i.e. on regulation, procurement, social and political processes, standards, criteria, success factors, etc.), new expectations were formed and new networks of CCA experts and advocates gained shape – such as politicians, climate activists, consultants, non-incumbent energy companies, trade associations, etc. An interesting example concerns the CCA 'inventor' and CCA evangelist Paul Fenn, who apart from training and 'spin-offing' other advocates and CCA experts, founded Local Power Inc., a consultancy company helping cities and regions develop more sophisticated versions of the CCA California model. These actors – territorially embedded yet spatially mobile – increasingly started to snowball and play an active role moving green and localized generation CCA models to other geographies within California and beyond.

3.3.4 The Spread of CCA in California and Beyond

Following Marin, Sonoma County was second to launch a CCA programme in California, followed by the city of Lancaster (Los Angeles County). At the time of

---

[6] At the time of this writing, San Francisco's CCA programme has already started the delivery of clean power to San Francisco.

this writing, several other counties and municipalities across California were in the process of launching/planning CCA programmes, inspired by the so-called California model. As expressed by a climate advocate in Sonoma during an interview,

there has been a lot of crosspollination between what happened in Marin and Sonoma, discussions about what kind of CCA [to implement], etc. . . . Marin was very generous and their elected staff came around all the time to tell the story in meetings, chair discussion sessions . . . and we used practically the same consultants. . . . I used myself to share a job at Marin with the former director of Marin Clean Energy.

As CCA concepts and practices became progressively legitimated and transformed in California, many actors became non-state intermediaries, actively and purposefully moving the concept across space and initiating new rounds of mobility. New consultancy companies were founded and became specialized, supporting municipalities in CCA processes (e.g. for legal and technical advice, energy modelling and forecasting, writing legal documents, public consultation, etc.). Local not-for-profit advocacy organizations gained constituency and financial resources to champion the concept across California and beyond – e.g. through the participation and organization of symposiums and conferences, public discussion sessions, etc. Some of them became sponsored by industry players with interest in the concept, such as electricity-generation companies (e.g. Shell). One example is LEAN Energy, which presented insights on the Californian CCA experience in the 2016 Paris COP meeting, together with a political representative from Marin. Other early advocates championing CCA in Marin and Sonoma started working for public energy agencies in California, while some became active in NGOs like the Sierra Club, thus influencing state and national energy policies and CCA diffusion through these channels.

Since 2011 and following the developments in California, transformed CCA concepts started 'moving back' eastward in the United States (see Figure 3.1), permeating energy regulations and practices in other states in different ways.[7] Some municipalities in pioneer CCA states (like Massachusetts and Ohio) started using CCA to offer long-term green energy options besides short-term savings. In Illinois, benefiting from an already largely liberalized electricity market, hundreds of local jurisdictions approved CCA – yet not always with an eye to procure/ generate cleaner electricity; municipalities like Chicago adopted CCA but moved back to the incumbent utility as contracts ended and retail prices decreased. In the state of New York, CCA is being deployed as a platform for distributed energy and the promotion of microgrids in Westchester County. Here again, CCA's concepts

---

[7] The analysis of CCA anchoring processes in these states is outside of the scope of this chapter.

and practices are being realigned with the state's strategies for energy resilience (in the aftermath of Hurricane Sandy) and with an eye to accelerate a transition towards greener and localized electricity portfolios. Paul Fenn wrote the CCA bill for New York, and the consultancy company of a former co-worker in California is advising Westchester County on electricity procurement.

3.4 Conclusions and Implications

This chapter discussed the relevance of a spatially informed perspective to understand where, how and through which channels (climate-related) experiments can progress and influence established socio-technical systems. Building on recent efforts to spatialize the local-global niche development model (e.g. Sengers and Raven, 2015), this chapter proposed a framework that looks at the diffusion of experiments as spatialized circuits of anchoring and mobility, inspired by notions put forward in the field of economic geography (e.g. De Propris and Crevoisier, 2009; Vale and Carvalho, 2013). This framework was illustrated with the case of CCA in the United States.

A number of implications from this analysis can be derived for the study of socio-technical transitions and the embedding of climate governance innovations. First, complementary to other notions of anchoring focusing on the linking patterns between niche and regime levels (Elzen et al., 2012), the notion of anchoring in this chapter highlights the socio-spatial embedding of experimentation – that is, the contexts and features of the multiple places where the socio-cognitive frames and the shared rules of global niches are formed (Geels and Raven, 2006). As the CCA anchoring processes in Marin vis-à-vis San Francisco illustrate, their concrete social, ecological and political features contribute in explaining why some places succeed (or fail) to adopt new alternatives and to make experiments gain traction towards forming global niches. Additional research is certainly needed to generalize these conclusions and understand these processes further; yet, recent studies for other contexts and innovation practices have been reaching similar conclusions (e.g. Sengers and Raven, 2015). Moreover, building on this perspective, further research could, for example, combine spatial and temporal dimensions of embedding to understand how the advantages of different types of places to foster experimentation may change along the lengthy journeys of niche formation and socio-technical change.

Second, the evidence in this chapter suggests that new concepts, visions and climate governance practices may not just 'aggregate' (Geels and Deuten, 2006): they can change substantially along their journeys of anchoring and mobility, as illustrated by the evolution of CCA from cost savings (Massachusetts) to green portfolios (Marin) to the localization of energy (San Francisco and beyond). The visions behind experiments become negotiated and recombined in place before they become mobile again, linked with concrete challenges of the place, actors' strategies and the interaction with other ongoing experiments (e.g. deregulation fiasco in California, energy localization and resilience strategies in New York). This localized crosspollination of experiments' concepts is in line with the findings of Seyfang et al. (2014) for grassroots sustainable energy niches in the United Kingdom and contributes to explain the mutation of concepts along experiments' spatial journeys.

Third, and linked with the previous evidence, the chapter highlights the frequent movement in space of concepts, rules and expectations, purposely channelled by actors that are, at the same time, territorially embedded and spatially mobile. The early mobility of CCA has been performed by a single mobile advocate, expert and initiator who physically moved across locations. Later intermediary actors have been formed to continue mobility and anchoring. Due to the tacit nature of some knowledge pieces involved with CCA (e.g. writing legal documents, steering political processes), there has been a strong overlap between actors involved in local processes and intermediaries – currently, a large share of the CCA intermediaries operating across the United States have been hands-on involved in the experiment in Marin. This suggests that, in some situations, effectively anchoring and moving experiments across space (and making them gain traction) require mobile intermediaries operating at both project and niche levels to reinterpret codified lessons and best practice guidebooks.

Overall, and in line with recent developments in the field of the geography of sustainability transitions (e.g. Truffer et al., 2015), this chapter suggests that building on more geographically sensitive frameworks may open new avenues for understanding which types of places are more relevant for the search and diffusion of new climate governance alternatives (as well as which ones may hamper it) and to identify the geographies of the actors and practices through which new concepts are transported beyond original sites to more effectively provoke change in socio-technical systems and climate-related governance.

Acknowledgements

The evidence in this chapter draws from a research project funded by Enel Foundation ('Energy Transitions in Cities: Lifestyle, Experimentation and Change', 2012–2014). We are particularly grateful to Ann Hancock, Leslie Alden and Paul Fenn for sharing their deep knowledge on CCA with us; to Erwin van Tuijl for co-conducting fieldwork in San Francisco; and to Catarina Maia and Diogo Ribeiro for drawing the chapter's maps. Luís Carvalho also acknowledges the financial support from the Portuguese Foundation for Science and Technology – FCT (BPD/103707/2014).

References

Asmus, P. (2009). Introduction to energy in California. In Phyllis M. Faber and Pavlik. B., *California Natural History Guides*. Berkeley: University of California Press.

Berkhout, F., Verbong, G., Wieczorek, A. J., Raven, R., Lebel, L., and Bai, X. (2010). Sustainability experiments in Asia: Innovations shaping alternative development pathways? *Environmental Science & Policy*, 13(4), 261–271.

Bessant, J., and Rush, H. (1995). Building bridges for innovation: The role of consultants in technology transfer. *Research Policy*, 24(1), 97–114.

Binz, C., Truffer, B., and Coenen, L. (2014). Why space matters in technological innovation systems – Mapping global knowledge dynamics of membrane bioreactor technology. *Research Policy*, 43(1), 138–155.

Binz, C., Truffer, B., and Coenen, L. (2016). Path creation as a process of resource alignment and anchoring: Industry formation for on-site water recycling in Beijing. *Economic Geography*, 92(2), 172–200.

Bulkeley, H., and Castán Broto, V. (2013). Government by experiment? Global cities and the governing of climate change. *Transactions of the Institute of British Geographers*, 38(3), 361–375.

California Energy Commission (2006). *Community Choice Aggregation*. Sacramento: Local Government Commission, California Energy Commission.

Carvalho, L. (2015). Smart cities from scratch? A socio-technical perspective. *Cambridge Journal of Regions, Economy and Society*, 8(1), 43–60.

Carvalho, L., Lazzerini, I., and van Tuijl, E. (2014a). Seventh Case Study: San Francisco. Working paper 16/2014, Enel Foundation, Rome.

Carvalho, L., Mingardo, G., and Van Haaren, J. (2012). Green urban transport policies and cleantech innovations: Evidence from Curitiba, Göteborg and Hamburg. *European Planning Studies*, 20(3), 375–396.

Carvalho, L., Santos, I. P., and Van Winden, W. (2014b). Knowledge spaces and places: From the perspective of a 'born-global' start-up in the field of urban technology. *Expert Systems with Applications*, 41(12), 5647–5655.

Centre for Climate Protection (12 February 2016). Active Community Choice Agencies and in the Process of Formation in California. Data sheet. Sonoma County, CA: Centre for Climate Protection.

Coenen, L., Benneworth, P., and Truffer, B. (2012). Toward a spatial perspective on sustainability transitions. *Research Policy*, 41(6), 968–979.

Coenen, L., Raven, R., and Verbong, G. (2010). Local niche experimentation in energy transitions: A theoretical and empirical exploration of proximity advantages and disadvantages. *Technology in Society*, 32(4), 295–302.

Crevoisier, O., and Jeannerat, H. (2009). Territorial knowledge dynamics: From the proximity paradigm to multi-location milieus. *European Planning Studies*, 17(8), 1223–1241.

De Propris, L., and Crevoisier, O. (2011). From regional anchors to anchoring. In Cooke, P., *Handbook of Regional Innovation and Growth* (pp. 167–177). Cheltenham: Edward Elgar.

Elzen, B., van Mierlo, B., and Leeuwis, C. (2012). Anchoring of innovations: Assessing Dutch efforts to harvest energy from glasshouses. *Environmental Innovation and Societal Transitions*, 5, 1–18.

Faulkner, K. (2010). Community Aggregation Choice in California. Unpublished paper, Berkeley: University of California.

Geels, F. (2002). Technological transitions as evolutionary reconfiguration processes: A multi-level perspective and a case-study. *Research Policy*, 31(8), 1257–1274.

Geels, F., and Deuten, J. (2006). Local and global dynamics in technological development: A socio-cognitive perspective on knowledge flows and lessons from reinforced concrete. *Science and Public Policy*, 33(4), 265–275.

Geels, F., and Raven, R. (2006). Non-linearity and expectations in niche-development trajectories: Ups and downs in Dutch biogas development (1973–2003). *Technology Analysis & Strategic Management*, 18(3–4), 375–392.

Hansen, T., and Coenen, L. (2015). The geography of sustainability transitions: Review, synthesis and reflections on an emergent research field. *Environmental Innovation and Societal Transitions*, 17, 92–109.

Hoogma, R., Kemp, R., Schot, J., and Truffer, B. (2002). *Experimenting for Sustainable Transport: The Approach of Strategic Niche Management*. London: Spon Press.

Kemp, R., Schot, J., and Hoogma, R. (1998). Regime shifts to sustainability through processes of niche formation: The approach of strategic niche management. *Technology Analysis & Strategic Management*, 10(2), 175–198.

Kivimaa, P. (2014). Government-affiliated intermediary organisations as actors in system-level transitions. *Research Policy*, 43(8), 1370–1380.

Lagos, M., and Baker, D. (12 November 2013), CleanPowerSF remains mired in politics. *San Francisco Chronicle*.

LEAN Energy (n.d.). CCA by State. Accessed 3 March 2016, www.leanenergyus.org/cca-by-state/

Local Power (2016). Community Choice Aggregation. Accessed 10 March 2016, www.localpower.com

NCSL (National Conference of State Legislatures) (December 2015). Community Choice Aggregation Policies. Accessed 3 March 2016, www.ncsl.org/research/energy/community-choice-aggregation.aspx

NREL (National Renewable Energy Laboratory) (2015). Status and Trends in the U.S. *Voluntary Green Power Market*. Report prepared under task No. SA15.0900. Golden, CO: NREL.

Raven, R., Schot, J., and Berkhout, F. (2012). Space and scale in socio-technical transitions. *Environmental Innovation and Societal Transitions*, 4, 63–78.

Sane Society (2012). Paul Fenn: Origins of Community Choice Aggregation. Online interview, YouTube, posted on 18 November 2012.

Schot, J., and Geels, F. W. (2008). Strategic niche management and sustainable innovation journeys: Theory, findings, research agenda, and policy. *Technology Analysis & Strategic Management*, 20(5), 537–554.

Sengers, F., and Raven, R. (2015). Toward a spatial perspective on niche development: The case of Bus Rapid Transit. *Environmental Innovation and Societal Transitions*, 17, 166–182.

Seyfang, G., Hielscher, S., Hargreaves, T., Martiskainen, M., and Smith, A. (2014). A grassroots sustainable energy niche? Reflections on community energy in the UK. *Environmental Innovation and Societal Transitions*, 13, 21–44.

Smith, A., and Raven, R. (2012). What is protective space? Reconsidering niches in transitions to sustainability. *Research Policy*, 41(6), 1025–1036.

Truffer, B., and Coenen, L. (2012). Environmental innovation and sustainability transitions in regional studies. *Regional Studies*, 46(1), 1–21.

Truffer, B., Murphy, J. T., and Raven, R. (2015). The geography of sustainability transitions: Contours of an emerging theme. *Environmental Innovation and Societal Transitions*, 17, 63–72.

US Department of Energy (28 January 2016). Green Power Markets – Community Choice Aggregation. Accessed 3 March 2016, https://www.nrel.gov/docs/fy18osti/70174.pdf

Vale, M., and Carvalho, L. (2013). Knowledge networks and processes of anchoring in Portuguese biotechnology. *Regional Studies*, 47(7), 1018–1033.

Wieczorek, A. J., Raven, R., and Berkhout, F. (2015). Transnational linkages in sustainability experiments: A typology and the case of solar photovoltaic energy in India. *Environmental Innovation and Societal Transitions*, 17, 149–165.

4

Realigning Circulations

How Urban Climate Change Experiments Gain Traction

VANESA CASTÁN BROTO AND HARRIET BULKELEY

4.1 Introduction

Experiments have gained currency in the study of the politics of climate change generally (Farrelly and Brown, 2011), but also in relation to how those politics play out in urban environments (Bulkeley and Castán Broto, 2013; Bulkeley, Castán Broto and Edwards, 2014; Karvonen and van Heur, 2014). Experiments have a key role in facilitating technological and social change shaping, for example, particular configurations of urbanism, whether this is through their relation with multiple urban sites or their emergence within a particular urban context. Overall, experiments emerge as a means through which to govern the city, in a context in which resource security concerns and climate imperatives direct attention to the governing of carbon and resource flows (see also, Hodson and Marvin, 2014).

Studies of experiments have helped develop a positive outlook on carbon governance, as an alternative to the international climate regime whose sluggish development and lack of local relevance has led to a growing sense of disenchantment (Hoffmann, 2011). Experiments can in part be explained by the proliferation of actors intervening in an uncertain environment (Biermann and Pattberg, 2012); yet, experiments are not always positive. There is a fundamental ambiguity about experiments and how they are linked to the configuration of spaces of authority, with the creation of carbon subjects and their consequent intervention in the consolidation of new forms of hegemonic power. However, they also constitute a challenge to existing configurations of global governance, and hence, they relate to the opening of spaces for practicing a politics of hope. There is an argument for experiments as demonstrating a current of optimism that believes in the possibility for intervention in global climate politics, which is often performed at the urban level (see also, van Asselt, Huitema and Jordan, Chapter 2). In this space of ambiguity, the growing interest in experiments has been followed by an equally passionate critique of their limitations, particularly in relation to the possibility of creating systemic change. This is both in terms of the extent to which experiments are able to generate new technological regimes and whether they challenge particular forms of urbanism at the city level (Evans, Karvonnen and Raven, 2016; see also, Karvonen, Chapter 11). This critique is one addressed in this book and, hence, the focus of this chapter.

The argument we wish to advance is that, to understand the power of experiments, we need to move away from explanations of systemic change that emphasise the possibility of scaling them up in a controlled or extractive sense. We advocate instead an engagement with the ways in which experimentation might be able to instantiate urban low-carbon transitions through the reconfiguration of urban infrastructure regimes and the possibilities they engender for changing their composition. Here change emerges in relation to the circulation of experiments in, through and of a particular urban milieu. We seek to unpack this notion of circulation, and why it matters, as a means to explain the role of experiments in wider processes of systemic change. The chapter thus explores the contemporary understanding of scaling up as a means to resolve governance anxieties. Then it turns to reflecting on what it means to conceptualise transitions as reconfigurations of urban infrastructure regimes. In this context, circulation emerges as a key concept which addresses how experiments can lead to transformative action. We conclude by mobilising an earlier set of cases of experiments and reflecting upon the forms of circulations that emerged in each of them.

4.2 Scaling Up: A Contemporary Anxiety in Innovation Processes

The notion of 'scaling up' reflects a series of concerns about climate change governance. It is often common to speak of scaling up in terms of moving from a concrete initiative, such as a piece of regulation, to broader impacts at a more general level (e.g. Nilsson and Persson, 2012). As most straightforwardly imagined, the idea behind the term 'scaling up' is that transformative changes require a change of scale, often reaching larger levels of change. The term uses the idea of scale in its everyday sense and, rather than the often quoted sustainability principle 'small is beautiful', it seems to advocate that 'big is necessary' for the kinds of changes required either because of their planetary proportions or because of the urgency of action. Such a reading of scale has been problematised in environmental governance debates (Bulkeley, 2005). Scale, or any spatial order for that matter, does not exist per se but rather is constituted through the socio-material assemblages and power relations through which governing is accomplished (Allen, 2004). As a result, scale becomes reconfigured in every attempt at creating diverse forms of governing (Reed and Bruyneel, 2010). Evidence has progressively accumulated about the multi-scalar nature of environmental problems and the multifarious ways in which scale is approached (Gunningham, 2009; Reed and Bruyneel, 2010). Yet, the notion of scaling up as a one-way transfer of an initiative or experiment from a smaller arena to a larger one persists, not just in terms of replicating and diffusing experiments such that they become commonplace but also through notions of increasing their reach and integrating local action in processes that are perceived to occur at a broader level.

The scalability of experiments has thus become a key characteristic which is used to, in some sense, measure their potential to create change at a systemic level. We see scaling up emerging as the ideal through which transformative change can be realised in relation to two different understandings of change of the city: one which relates to ideas of governing change in the city, particularly through spatial planning, and another which emerges from ideas of diffusion of innovations.

In relation to the first perspective, the popularity of the notion of scaling up needs to be understood in the context of broader changes in planning theory and practice in which the possibility of planning 'at scale' has been replaced by more incremental notions of how cities are made. While the idea of a systematic, centralised and controlled approach to planning the city used to dominate this arena, planners talk now of a communicative turn in which there is an abandonment of map-based or scientific certainties in favour of discursive forms of rationality (Tewdwr-Jones and Allmendinger, 1998; Healey, 2003; Gunningham, 2009; Innes and Booher, 2010), as inspired by the German philosopher Jürgen Habermas. While having a strong philosophical basis, scholars of collaborative planning have built on their own empirical observations about how planning occurs (Healey, 2003). Despite critiques that emphasise the extent to which collaborative planning can truly address power relations and ensure outcome delivery (e.g. Tewdwr-Jones and Allmendinger, 1998; Clifford and Tewdwr-Jones, 2013), forms of communicative planning have become standard to address environmental transformations in the context of complexity (Innes and Booher, 2010). Whatever the flavour of collaborative or communicative planning, there is a sense of abandoning certainties in urban development and the extent to which these new planning notions fit into neoliberal discourses.

In response to these developments there has been an emerging nostalgic impulse to plan in what are thought to be more traditional ways, putting the state at the centre of a process of authority building (e.g. Olesen and Richardson, 2012). Here, scaling up is used as a means of reclaiming urban planning and governing by control, in which predesignated experts manage uncertainty and use different computational mechanisms to deliver adaptable planning (Verebes, 2013). The need to achieve scale is often linked to a return to conventional planning processes, in which contemporary thinking about collaborative planning is put to work to enrol relevant actors in predefined processes of spatial ordering (Turner, 2016).

Scaling up is an imperative necessary to address the 'institutional void' (Hajer, 2003), the fragmentation of authority that has followed both the rise of neoliberal discourse and the growing scepticism about expert-led processes of decision-making. Even if planning is considered as a more iterative, adaptable form, scaling up is a means for ordering those realms of authority in which experimentation is possible as localised or smaller levels of action, while governments play a key role in organising, systematising and reproducing such localised efforts. Equally, scaling up is a means to deal with the materiality of processes and practices of urban planning. Take for example master planning, a quintessential practice in urban planning, which is now enjoying a new lease of life by virtue of new planning practices facilitated by big data and social media (Verebes, 2013). Whatever its results, master planning is full of experimentation, a series of insertions around a particular problem. Yet, the outputs of master planning are most often presented as a reified series of maps which designate both spaces of state authority and possible futures. The map (or the regulation, or the state-led programme) becomes then the principal means whereby initial experiments are scaled up. Yet, the desire for systematisation is not equivalent to the actual material reproduction of an experiment or its principles at higher scales which are often seen as falling short of such ideals. Furthermore, our work suggests that this is a limited way to understand the role of experimentation, and one that responds more to a concern with the need to deliver effective action or to control certain forms of hegemonic power emerging in the neoliberal context, than to an actual understanding of how systemic change towards low-carbon, resilient cities can take place. Where urban planning is being mobilised as a means through which climate change is being addressed in cities, it provides a more ambiguous arena through which experimentation is both strategically inserted and enrolled into different programmes, brought into dialogue with existing planning practices and leveraged as a means of lending support to both incumbent actors and their critics (Carmin, Anguelovski and Roberts, 2012; Chu, 2016).

The second perspective through which scaling up has come to be seen as an important means to deliver transitions derives from market-oriented approaches to innovation diffusion (e.g. Loorbach and Wijsman, 2013). Traditionally, innovation diffusion ideas emerged in relation to Schumpeterian ideas of innovation as a discrete but collective outcome distinct from inventions (Robertson, 1967). From this perspective, oriented towards businesses and shaped by conventional notions of rational economic actors, the most difficult issue is not generating innovation but ensuring its diffusion. This kind of thinking has evolved, but the fundamental ideals of achieving scale have remained at the core of market-oriented ideas of innovation. We can observe that transitions studies – a body of work with which we have engaged exhaustively in our previous work – remains very much oriented towards understanding processes of structuration to foster a transition (Geels and Schot, 2007). Most often, studies of transitions using the multilevel perspective (Geels, 2005) or focusing on community-based innovation (Hargreaves et al., 2013) assume a theory of change in which the emergence of innovation in niches spreads further. The issue here becomes the pathways followed so that a certain alignment between localised innovations and structural variables can facilitate a change of regime (Geels and Schot, 2007). The theory identifies opportunities for diffusion within a larger institutional framework so that innovators work alongside publics to make a transition possible.

Notions of experimentation are at the centre of the transition process, as often described within the subfield concerned with transitions management, which aims to link academic and operational thinking to put transitions in practice. Experiments constitute an initial step towards transitions that should be broadened, deepened and ultimately scaled up through an orchestrated process of transitions management (Loorbach, 2010). Here, scaling up refers to the process of moving from within experimentation in niches to their incorporation in regimes. The overarching preoccupation is with the visions, institutions and coalitions through which such processes of scaling can take place. Ideas of directionality and increasing scope are central to the thinking of how experiments can actualise systemic change and to the practice of transition management as a means of intervening in the urban arena.

In contrast, the perspective on experimentation as a mode of governing the urban milieu that we have developed (Bulkeley et al., 2014) departs from these perspectives in at least four ways. First, neither the urban planning nor the market-oriented perspective on experiments and change reflects our understanding of experiments as emerging from urban assemblages, that is, as created purposively within a series of material and discursive resources that are brought together through a process of experiment-making. Second, emphasis on scaling up relates to static notions of scale, whether this is a scalar space in the case of urban planning or to heuristic notions that establish an artificial separation between hegemonic regimes and the spaces of experimentation. Third, both approaches suggest that there is a possibility to define and enforce definitive realms of authority, as if authority was not constantly being construed within particular realms of discursive and material action – in short as if 'power' operates outside the urban milieu and can be applied to it, rather than being emergent from the ways in which it is constituted. And fourth, there is a discounting of the material unruliness of the urban regime, either because it is thought to be tamed through planning practices or because it is celebrated as the source of diversity which automatically leads to innovation but which is then held at bay once natural selection has taken place. Overall, we find that the predominant discourses of scaling up relate to a passive role for experiments as configuring change, as if the actual process of change was not already embedded in their making, but required coordinated efforts to make it visible. This is not to suggest that systemic change is unnecessary. Far from it. This is to suggest that experimentation as a mode of governing already contains the means through which systemic change can be afforded, as we explain in the following section.

4.3 The Reconfiguration of Urban Infrastructure Regimes

Rather than use the term 'scaling up', we propose to focus on the work of experimentation and how it creates the apertures for new circulations that can open up socio-material configurations. First of all, there is a need to reflect on the policy context of the experiment and problematise the notion that the governance landscape is in any sense unified or integrated, so that it is doubtful whether a process of scaling up – which takes for granted that the experiment is located in such a unified space – as currently imagined is even possible. Instead, from our perspective, experiments emerge within an ecology of interventions across a landscape that is always multi-scalar in its configuration (McGuirk, Bulkeley and Dowling, 2014), and these experiments relate to each other in multiple ways which can occasionally be systematised in specific and tailored strategic pro-grammes. In relation to urban planning, it is not that action builds on previous experiments, so that planning becomes a more authentic means of action which can be coordinated, but rather, that previous experiments are discursively enlisted into the actual strategic project of making a plan. A similar process of market innovation occurs.

In an urban context, experimentation is co-ordinated and strategically invested to achieve some form of transformation in urban infrastructure regimes. This kind of thinking about transformation engages both with the socio-technical and terri-torial dimensions of transitions (Hodson et al., 2012). Urban infrastructure regimes refer to the socio-technical conglomerate that enables the use of technology in a particular social, cultural and technological context. Urban infrastructure regimes challenge static notions of infrastructure as purely material, emphasizing how such infrastructure materials are inseparable from a set of norms, conventions and regulations that define the role of infrastructures and the behaviours around them. This notion also makes explicit the territorial dimensions of socio-technical configurations, particularly in relation to their dependence on the metabolisation and redirection of flows of resources (Monstadt, 2009). Urban infrastructure regimes are cohesive enough that they are characterised by resistance to change – that is, they obdurate (Hommels, 2005). Yet, the elements of urban infrastructure regimes are dynamic and subject to ongoing work through processes of maintenance (Graham and Thrift, 2007; Castán Broto and Bulkeley, 2013). Change is thus continuous, as it is predicated on the operation of urban infrastructure regimes. Indeed, regimes may evolve along specific pathways. However, fundamental transformations of those regimes are rare, even when radical innovations are incorporated into those regimes. While this notion of regime emerges in dialogue with the socio-technical regimes at the core of the multilevel perspective, the emphasis here is on the contingent, immanent character of those regimes and how they are shaped by specific contexts of urban development.

Experiments are fundamental to transformations because of the work they perform in creating apertures through which alternatives to urban infrastructure regimes can emerge. This 'opening' of urban infrastructural regimes takes place through the creation of new spaces of authority; the development of calculative means to facilitate the intervention; the inscription of new practices in material devices; the engagement with unruly materiality through demonstrations; and the questioning of established ways of acting within particular devices. Any process of controlling or managing the kinds of change that emerge from experiments is always taking place alongside the ongoing dynamics of change fostered by the experimental intervention itself. In short, experimentation brings about an *unruly* dynamic of new configurations and potential circulations which may be more or less successful in opening up existing obdurate socio-material regimes. In this sense, the transformative potential of the experiment is always and already part of the intervention, and its capacity for transformation relates both to the 'stickiness' of the regime within which it is inserted and its immanent capacity to assemble, enrol and transform the socio-technical configurations of which it is a part.

Thus, the question that we need to ask is not how experiments can be scaled up but rather, how experiments mobilise so that transformative change can emerge. Mobilisation refers to a process whereby the material discourses and resources that make the experiment are brought together to facilitate action. Rather than this involving 'power over' the urban milieu, it is born from an immanent and relational power of experimentation as a mode of governing. This capacity to facilitate action has different dimensions in terms of directing, accelerating and circulating. Of these, we focus on circulation because circulation is central to the configuration of urban infrastructure regimes. In the following section we attempt an initial conceptualisation of the notion of circulation in relation to the role of experiments in urban transformations.

4.4 Realigning Circulations

Circulation thus emerges as a key concept to understand the role of climate change experiments in fostering actual transformations of urban infrastructure regimes.

How might this be understood? Within the urban political ecology imaginary circulation has been tied up with ideas of metabolism, linking circulation with the actual movement of substances through infrastructure. Yet, when thinking of circulation in relation to experiments, and even when doing so from a decidedly materialistic point of view, circulation is not limited exclusively to the movement of things. Our own work relates to three different, but interrelated understandings of circulation.

First, circulation can be thought of as the actual movement of things in relation to different assemblages that constitute the urban infrastructure regime. This is an idea that emerges from within an actor-network theory (ANT) which both situates things within assemblages and recognises the efforts at integrating things through a progressive process of enrolment and translation (Murdoch, 1998, 1997; Latour, 2005). From within an ANT perspective, this movement is not a physical change of location in the metabolic sense, but rather, it is a change of the set of relations that integrate a 'thing' (be it a resource, technology, discourse, belief) within a particular regime. The notion of circulation, however, implies an ordering in the realigning of relations. An experiment thus can be mobilised to configure new sets of relations around urban infrastructure regimes. For example, a solar lamp sold to residents in informal settlements will require building new understandings of energy use, business models and material connections that situate that lamp within a particular style of construction and habitation. In doing so, solar lamps do not only establish a new set of connections (so that they can be used in that particular context) but also open questions about the parameters established in a given urban infrastructure regime, in particular, the structural factors that make it possible. The exact shift in relations will depend on the particular context in which they take place, and hence, the impact of the experiment is contingent on the specific situation in which it is assembled, maintained and lived.

Second, circulation can be understood not in relation to what moves and the relations which are established, but in relation to the receptors or nodes that make that circulation possible. This is, in some sense, the counterpart of the perspective proposed earlier. If the previously proposed perspective emphasises 'things' that circulate, then this second perspective emphasises 'things' that make circulation possible. This is something that resonates with perspectives on policy mobilities which already influenced our previous work on experiments (Bulkeley et al., 2014). In this literature, the focus is on the assemblage of policies, and hence, what is mobilised is, particularly, knowledge (McCann, 2008, 2011; McCann and Ward, 2012). This is a literature that emphasises policies as emerging within specific settings of relations. However, key to the notion of circulation is the idea that such policies move from one point to another as '[m]obilities are, nonetheless, tied to and facilitated by various 'moorings', organizing nodes, or fixed infrastructures' (Temenos and McCann, 2013: 346). This is not akin to thinking of scalar locations in the map across which policies move, but rather, the movement is from one set of relations to another, with the difference that such sets are concentrated around a node. The key revealing issue is that, while policies in the policy mobilities literature remain relational, processes of circulation make it possible for specific actors to reimagine them within specific contexts of action. Roy and Ong (2011) have put experiments, for example, at the centre of the work of middling bureaucrats who depend on the appropriation of global discourses within their specific contexts of operation. While building on notions of assemblage, however, the literature on policy mobilities emphasises the discursive aspects of mobilisation, but a similar approach can be used to think of the materiality of experiments and how certain physical and technological characteristics travel alongside ideas and beliefs. The capacity both to translate, discursively, and manipulate, materially, is simultaneously crucial for the circulation of experiments, although policy experiments may tend to emphasise their discursive components. Thus, the mobilisation of experiments through their circulation from one node to another is related to change in so far as it may challenge the position of certain nodes and arrange for new ones, not around what moves but around what allows that movement. For example, the mobilisation of experimental templates for sustainable housing in Mexico and their subsequent translation into a national programme for the delivery of ecotechnologies in low-income settlements led to a fundamental transformation in the housing industry, because it supported the development of new financial mechanisms to support the incorporation of eco-technologies in low-income houses through an instrument called the Green Mortgage. However, although the experiment was initiated with the development of housing templates, mobilisation only happened with the changes that took place among developers and other key industry actors during the adoption of such templates, rather than with a transformation of residents' practices or institutions' models of delivering low-income housing. The experiment was not necessarily repositioned, but it transformed the possibilities for the actors ('nodes') who could intervene in delivering transformative change.

Finally, the third component of circulation relates to the amplification, reach and dissemination of experiments. One key step for an experiment to reverberate across the system is through the creation of publics receptive to such experiments. Thus, unlike ideas of technology diffusion that emphasise awareness and education of consumers, circulation tackles head-on the idea that creating the means of governing through experiments entails the creation of subjects who are able to act in relation to that experiment. However, there is a difference between a localised process of subjectification which occurs in relation to a single experiment (e.g. an eco-community redefining its habits in relation to the architecture or design of the place) and the kind of public that is created when a design opens up potential futures and moves other people to try it out and continue experimenting. This marks the difference between experimentation and a process of experiment-led transformation, in which there is both a material transformation and a re-imagination of the processes that the experiment was intended to change. Amplification is not simply repetition of the experiment, but rather, amplification is the action that follows the realisation of alternative futures as emerging from experiments. In transition studies, scholars have spoken of 'pre-figurations' of alternative socio-technical systems (Turnheim et al., 2015). However, through amplification the experiment is not a prefiguration because the transformative power of the experiment relates to the possibilities it opens for actors to actively create new forms of identity that can deliver alternative futures. Prefigurations can mobilise actors' expectations, but the process of amplification is one in which such expectations are actively reconfigured into new forms of authority. Unlike the two previous understandings of circulation, this is one understanding that engages with the most discursive aspects of experiments, and how they are tied to particular modes of understanding the world.

4.5 Reflections on Circulation Emerging from Case Study Comparisons

Our book (with Gareth Edwards) *An Urban Politics of Climate Change* (Bulkeley et al., 2014) presents the case for understanding experimentation as a means of climate change governance in urban areas. The book concludes with a reflection about the transformative potential of experiments. To support the analysis, we present eight case studies on how climate change experiments are made, maintained and lived in a particular urban setting. Each case study is located in a city which is, in some way, global. Although there are precedents in some cases, the eight case studies focus on initiatives for climate change that emerged after the ratification of the Kyoto Protocol in 2005. The case studies are:

- Towards Zero Carbon (T-Zed) is a housing development on the outskirts of Bangalore, India, which experimented with cutting edge technologies to deliver low-carbon housing for India's emerging middle class, detailing its emergence.
- The project ViDA (Vivienda de diseño ambiental) was an experiment of low-carbon housing for a low-income neighbourhood in Monterrey, Mexico, in which government institutions attempted to demonstrate the viability of delivering bioclimatic design at low cost.
- The Retrofit Philly 'Coolest Block' Contest was an experiment to retrofit housing for energy conservation in the U.S. city of Philadelphia. Supported by

DOW Chemicals, the contest brought public and private interests to create a competition for a retrofit or weatherization prize.

- The ceilings project in Mamre, Cape Town, was another experiment in housing retrofit for the provision of insulated ceiling installations in privately owned housing in very low-income areas. The project emerged from a partnership between Danish International Development Assistance (DANIDA) and was led by the city of Cape Town.
- The Solar Atlas and the Solardachbörse (Solar Roof Exchange) were two experiments in Berlin, Germany, which attempted to demonstrate the potential of solar technologies and facilitate their installation through making available technical knowledge about where these technologies could be installed.
- The solar hot water (SHW) experiments in São Paulo, Brazil, have been delivered in the social housing sector by the Housing and Urban Development Company of the state of São Paulo (CDHU), Brazil's largest housing agency.
- PowerSave, promoted through Friends of the Earth and Climateers, led by the World Wildlife Fund in the city of Hong Kong, China, are two parallel experiments designed to act on individual behaviour and energy use through contexts, networking and social media.
- The Low Carbon Zone in Brixton, London, is an experiment that combined the efforts of the Greater London Authority and local community organisations, including the emergent Transition Towns Brixton, to help intervene in different areas of energy provision and consumption in the city.

Taken together, these projects display different circulatory processes at play in climate change experimentation (Table 4.1). For example, they show that there is a range of actors working to gain authority over the climate change arena, and they do so by drawing on the resources and networks that they have at hand, sometimes creating unlikely alliances. If we focus on the ways these projects have transformed the landscape of climate change governance, we see that, while all of them had certain importance in terms of their prefigurative impact, not all of them led to broader reconfigurations of the urban infrastructure regime. Circulation as movement means that the 'kernel' of the experiment – the principle that inspires it (whether it is a specific technology or way of doing things) – moves successfully from one context to another. In the case of ViDA, there has been a true movement from a localised project to housing templates that have enabled broader changes in the construction and regulatory landscape. The Coolest Block does not appear to have a strong transformative impact, perhaps because its major impact was in terms of amplification, but the context did not serve to create new publics that would engage the experiment in new contexts. The cases of Mamre's ceilings experiment in Cape Town and Solar Water Heaters in São Paulo both demonstrate circulation as movement, in the sense

Table 4.1. *Circulatory processes in eight climate change experiments*

	Circulation as movement	Circulation as change of position	Circulation as amplification
T-Zed, Bangalore	Movement from housing templates to an eco-house regulation; requires realigning institutions	The example of T-Zed has worked to redefine the work of developers and consultants.	The example of T-Zed has created new publics for eco-housing.
ViDA, Monterrey	Limited transformative impact, in part because of its limited circulation across different actors and little building of additional publics		
Coolest Block, Philadelphia			
Ceilings, Cape Town	The circulation of insulation technologies has redefined the circuits of public health, although overall impact limited.		
Berlin, Solar City	Limited transformative impact, especially because the atlas could not circulate from one position to another, in terms of both who could use the technology and what it represented		
Solar Water Heaters, Sao Paulo	Circulation of solar technologies in different housing programmes; helping to configure them		
Power Save and Climateers, Hong Kong			Subjectification and amplification through participation in collective activities and sharing experiences
Low Carbon Zone, Brixton		Circulation of principles and their appropriation from one urban context to another	

that is a particular bounded technology that is mobilised across the city, and in the case of Brazil, beyond this particular city. The extent to which this circulation is eventually transformative will depend on the context of uptake, something that we discussed at length in the last chapter of our book.

Circulation as change of position involves experimental processes whereby the 'receptors' are changed, and henceforth they move onto making a different urban infrastructure regime. In the case of T-Zed, the experiment gained traction by redefining the terms of reference of professionals in the housing sector and creating a 'public' of high-income earners which are now demanding new standards in housing. While the first one represents a means of circulation, helping change the 'nodes' (here consultants, developers, architects), the latter form of change represents an amplification of the experiment (see Table 4.1). Both solar experiments in Berlin were designed to enable circulation as a change of position, directed again to change the receiving 'nodes' and configuring houses and professional practices to enable the normalization of solar technologies, but the experiment was not clearly successful in achieving this. Perhaps the clearest case is the Low Carbon Zones in Brixton, in which there was a profound transformation among the approaches of the public sector and the communities which overall move to a new set of low-carbon practices.

Finally, circulation by amplification implies the creation of new sets of publics, new arenas or spaces where new practices can emerge, hence leading to different urban infrastructure regimes. The active creation of a new demand for eco-inspired housing for high-income earners in India with projects like T-Zed is a clear case of amplification. In Hong Kong, the experiments Power Save and Climateers actively sought to enable new forms of subjectification whereby publics will take the initiative to demand and deliver low-carbon action, in an example that can be characterised above all the others, by a complete absence of the state. Overall these cases demonstrate the complex dimensions of circulation and the different ways in which experimentation could be transformative in terms of challenging the urban infrastructure regime.

4.6 Conclusion

In this paper, we have sought to question whether the idiom 'scaling up' is the most useful means through which to imagine, conceptualise and investigate the ways in which experiments take effect and have transformative consequences. We suggest this notion of scaling up as impact is linked both to a nostalgia for the possibilities of top-down planning and to broader ideas of diffusion of innovation. In both conceptualisations, we find scaling up as a limiting concept in terms of explaining the transformative power of experiments. This is because the transformative power of experiments relates to the possibilities they create within urban infrastructure regimes. This is not to do with the extent that external forces can ensure their diffusion, replication or scalability, but rather, it is to do with the experiments themselves and how they are inserted in a particular web of socio-technical relations. Transformations – while rare – relate to the possibilities to mobilise those experiments and establish new connections. There is also a question about what counts as transformation, and whether transformations can be really visible. Transformations can happen in many ways. There will always be a temptation to connect transformations with direct interventions, in which particular actors engage with mechanisms of control over urban infrastructure regimes, through regulation or other top-down interventions. In other cases, actors may be looking to accelerate the pace of change by throwing resources and efforts at a particular experiment. Perhaps more effort is needed to understand directing and accelerating mechanisms whereby experiments may lead to transformations. In this chapter, however, we have focused on the often overlooked mechanism of circulation, showing the nuanced ways in which those mechanisms are put into practice. We have proposed three components to understand the transformative power of circulation:

- Circulation as the actual movement of the experiment
- Circulation in terms of the origin and destination of the experiment – where from and where to it moves
- Circulation in terms of amplification and public reaction

This is not to say these are clearly isolated. Most often, different forms of circulation will happen simultaneously, reinforcing each other. However, as the analysis of experiments shows, some circulation aspects may be more visible than others depending on the experiment. We have started analysing this empirically with reference to previous case studies of experiments which were linked (or not) to actual transformations.

Overall, there is no one single way to link experimentation with transitions to sustainability or low-carbon transitions. However, it seems that experimentation may work best when it actively seeks to create the capacity for circulation from the outset, for example, by engaging with forms of socio-technical innovation that are relatively mobile, by reaching to different positions through which those circulate and by seeking the amplification of experimentation practices.

References

Biermann, F., and Pattberg, P. H. (2012). *Global Environmental Governance Reconsidered*. Cambridge, MA: MIT Press.

Bulkeley, H., and Castán Broto, V. (2013). Government by experiment? Global cities and the governing of climate change. *Transactions of the Institute of British Geographers*, 38(3), 361–375.

Bulkeley, H. A., Castán Broto, V., and Edwards, G. A. (2014). *An Urban Politics of Climate Change: Experimentation and the Governing of Socio-technical Transitions.* London: Routledge.

Carmin, J., Anguelovski, I., and Roberts, D. (2012). Urban climate adaptation in the global south planning in an emerging policy domain. *Journal of Planning Education and Research*, 32(1), 18–32.

Castán Broto, V., and Bulkeley, H. 2013. Maintaining climate change experiments: Urban political ecology and the everyday reconfiguration of urban infrastructure. *International Journal of Urban and Regional Research*, 37(6), 1934–1948.

Chu, E. (2016). The political economy of urban climate adaptation and development planning in Surat, India. *Environment and Planning C: Government and Policy*, 34(2), 281–298.

Clifford, B. P., and Tewdwr-Jones, M. (2013). *The Collaborating Planner?* Bristol: Policy Press.

Evans, J., Karvonnen, A., and Raven, R. (2016). *The Experimental City.* Abigndon, NY: Routledge.

Farrelly, M., and Brown, R. (2011). Rethinking urban water management: Experimentation as a way forward? *Global Environmental Change*, 21(2), 721–732.

Geels, F. W. (2005). *Technological Transitions and System Innovations: A Co-evolutionary and Socio-technical Analysis.* London, Amsterdam: Edward Elgar Publishing.

Geels, F. W., and Schot, J. (2007). Typology of sociotechnical transition pathways. *Research Policy*, 36(3), 399–417.

Graham, S., and Thrift, N. (2007). Out of order understanding repair and maintenance. *Theory, Culture & Society*, 24(3), 1–25.

Gunningham, N. (2009). The new collaborative environmental governance: The localization of regulation. *Journal of Law and Society*, 36(1), 145–166.

Hajer, M. (2003). Policy without polity? Policy analysis and the institutional void. *Policy Sciences,* 36(2), 175–195.

Hargreaves, T., Hielscher, S., Seyfab, G., and Smith, A. (2013). Grassroots innovations in community energy: The role of intermediaries in niche development. *Global Environmental Change*, 23(5), 868–880.

Healey, P. (2003). Collaborative planning in perspective. *Planning Theory*, 2(2), 101–123.

Hodson, M., and Marvin, S. (2014). *After Sustainable Cities?* London: Routledge.

Hodson, M., Marvin, S., Robinson, B., and Swilling, M. (2012). Reshaping urban infrastructure. *Journal of Industrial Ecology*, 16(6), 789–800.

Hoffmann, M. J. (2011). *Climate Governance at the Crossroads: Experimenting with a Global Response after Kyoto.* Oxford: Oxford University Press.

Hommels, A. (2005). *Unbuilding Cities: Obduracy in Urban Sociotechnical Change.* Cambridge, MA: MIT Press.

Innes, J. E., and Booher, D. E. (2010). *Planning with Complexity: An Introduction to Collaborative Rationality for Public Policy.* London: Routledge.

Karvonen, A., and van Heur, B. (2014). Urban laboratories: Experiments in reworking cities. *International Journal of Urban and Regional Research*, 38(2), 379–392.

Latour, B. (2005). *Reassembling the Social: An Introduction to Actor-Network-Theory.* Oxford: Oxford University Press.

Loorbach, D. (2010). Transition management for sustainable development: A prescriptive, complexity-based governance framework. *Governance*, 23(1), 161–183.

Loorbach, D., and Wijsman, K. (2013). Business transition management: Exploring a new role for business in sustainability transitions. *Journal of Cleaner Production*, 45, 20–28.

McCann, E. (2011). Urban policy mobilities and global circuits of knowledge: Toward a research agenda. *Annals of the Association of American Geographers*, 101(1), 107–130.

McCann, E., and Ward, K. (2012). Policy assemblages, mobilities and mutations: Toward a multidisciplinary conversation. *Political Studies Review*, 10(3), 325–332.

McCann, E. J. (2008). Expertise, truth, and urban policy mobilities: Global circuits of knowledge in the development of Vancouver, Canada's 'four pillar' drug strategy. *Environment and Planning A*, 40(4), 885–904.

McGuirk, P., Bulkeley, H., and Dowling, R. (2014). Practices, programs and projects of urban carbon governance: Perspectives from the Australian city. *Geoforum*, 52, 137–147.

Monstadt, J. (2009). Conceptualizing the political ecology of urban infrastructures: Insights from technology and urban studies. *Environment and Planning A*, 41(8), 1924–1942.

Murdoch, J. (1997). Inhuman/nonhuman/human: Actor-network theory and the prospects for a nondualistic and symmetrical perspective on nature and society. *Environment and Planning D: Society and Space*, 15(6), 731–756.

Murdoch, J. (1998). The spaces of actor-network theory. *Geoforum*, 29(4), 357–374.

Nilsson, M., and Persson, Å. (2012). Reprint of 'Can Earth system interactions be governed? Governance functions for linking climate change mitigation with land use, freshwater and biodiversity protection'. *Ecological Economics*, 81, 10–20.

Olesen, K., and Richardson, T. (2012). Strategic planning in transition: Contested rationalities and spatial logics in twenty-first century Danish planning experiments. *European Planning Studies*, 20, 1689–1703.

Reed, M. G., and Bruyneel, S. (2010). Rescaling environmental governance, rethinking the state: A three-dimensional review. *Progress in Human Geography*, 34(5), 646–653.

Robertson, T. S. (1967). The process of innovation and the diffusion of innovation. *Journal of Marketing*, 31(1), 14–19.

Roy, A., and Ong, A. (2011). *Worlding Cities: Asian Experiments and the Art of Being Global.* Chichester: John Wiley & Sons.

Temenos, C., and McCann, E. (2013). Geographies of policy mobilities. *Geography Compass*, 7(5), 344–357.

Tewdwr-Jones, M., and Allmendinger, P. (1998). Deconstructing communicative rationality: A critique of Habermasian collaborative planning. *Environment and Planning A*, 30(11), 1975–1989.

Turner, V. K. (2016). How do conventional master planning processes facilitate or constrain sustainable urbanism? An environmental management perspective. *Society & Natural Resources*, 29(12), 1483–1500.

Verebes, T. (2013). *Masterplanning the Adaptive City: Computational Urbanism in the Twenty-First Century.* London: Routledge.

5

Understanding Public Dialogue as an Embedded Democratic Innovation in UK Climate Governance

HELEN PALLETT

5.1 Introduction

For the last decade or more, there has been a large degree of consensus among policy actors, civil society groups and academics that the scale and scope of the transitions required to address the multi-faceted challenge of climate change necessitates engagement and dialogue with citizens (Pallett and Chilvers, 2013). Therefore, techniques for public participation and engagement have played a central role in climate governance, helping identify and describe the nature of the challenge and developing interventions to further climate change mitigation and adaptation measures, as well as challenging or holding to account the actions of governing institutions. As has been demonstrated in relation to environmental governance more broadly (Brown, 2009; Munton, 2003), public participation has been increasingly institutionalised in climate governance structures to the extent that it is difficult to identify decisions and interventions around which no public engagement has taken place (Pallett and Chilvers, 2013). Furthermore, this institutionalisation can be observed at multiple scales and in many domains, with public participation techniques being routinely adopted by departments and agencies of local, national and transnational government, as well as by non-governmental organizations (NGOs) and charities, businesses and scientific institutions such as universities and research councils.

To go beyond the conventional focus on the quality and impacts of individual participation or governance processes, and to engage with the incipient institutionalisation of participation, this analysis focuses on a particular technique of public participation and climate governance – namely the UK government's *public dialogues*, which it has been carrying out and promoting for more than a decade. There are multiple forms of experimentation at play in and around the practice of public participation. Perhaps most obviously, participation processes themselves have been characterised as experiments, in the sense of being tests of particular policies and knowledge claims. But these processes also play an important role in broader experiments in policy learning and the politics of ordering collective reality around climate change polices (cf. Turnheim, Kivimaa and Berkhout, Chapter 1).

This chapter explores these multiple experiments around institutionalised public participation processes using the concept of *democratic innovations*. This concept comes from the political science literature and has been used to refer to the creation and institutionalisation of new techniques for public participation (Smith, 2009). It also has resonances with related work in human geography and science and technology studies (STS) concerned with techniques of participation, which has taken more seriously the implications of analysing these practices as innovations or technologies (Peck and Theodore, 2015; Voß, 2016). This chapter develops the concept of democratic innovations further in reference to this work and to debates about experimentation in human geography, STS and the history of science, through the case study of public dialogue.

The UK government's public participation expert resource centre Sciencewise has been developing and promoting the technique of *public dialogue* since its creation in 2004. During this time the programme has supported more than fifty public dialogue processes in partnership with different government departments, agencies or research councils, more than one-third of which have directly engaged with climate governance. Some of the most prominent examples of climate-related public dialogues include the Big Energy Shift, the 2050 Pathways Dialogue and the Low Carbon Communities Challenge, all of which were co-sponsored by the Department for Energy and Climate Change (DECC), whilst other dialogues have included topics as diverse as bioenergy, climate change adaptation, geoengineering, flood preparedness and shale gas. The empirical material presented in this chapter comes from two phases of qualitative research on the Sciencewise programme. The first phase was a historical study of the emergence and changes in the Sciencewise programme 2000–2010 carried out with semi-structured interviews and document analysis in 2011 (see Pallett and Chilvers, 2013), and the second phase was an ethnography of the programme carried out throughout 2013 including participant observation, semi-structured interviews and document analysis (see Pallett, 2015).

This chapter is structured as follows. Section 5.2 considers the relevance of metaphors of experimentation to the study of democracy and participation. Building on this, Section 5.3 introduces and explores the analytical value of the concept of democratic innovations. Section 5.4 then describes and examines the case of Sciencewise's public dialogues as a democratic innovation. Finally, Section 5.5 considers the different ways in which these experiments have overflowed their bounds, with significant implications for climate governance and public engagement.

5.2 The Laboratory of Participation

The emergence of increasingly institutionalised and standardised modes of participation has been the subject of growing criticism from STS scholars. Bogner expresses this argument perhaps the most eloquently by describing the forms of participation currently adopted by governing institutions as taking the form of a laboratory experiment (Bogner, 2012). This metaphor highlights the level of control exerted through the orchestration of these processes, driven by the concurrent professionalization and commercialisation of participatory practices. The participation laboratory (participation process) is a closed environment which contains a particular set of apparatus (methods of participation) and which only some citizens are permitted to enter. Furthermore, the aims and the rules of the participation experiment are predetermined by the orchestrators of the experiment (the institutional commissioners and facilitators). Here, Bogner and other authors highlight the narrowing set of methods which are considered to be 'best practice' in public participation (Chilvers, 2008a; Cooke and Kothari, 2001). It has also been observed that this narrowing of methods has created an increasingly exclusive group of participation experts or mediators, with the power not only to define what constitutes good and bad participation but also to design and carry out participation processes and to speak on behalf of citizens in the context of science policy (Chilvers, 2008b; Gisler and Schicktanz, 2009; Osborne, 2004). Therefore, the sealed laboratory of participation has not only implications for how citizens are engaged and constructed but also what is considered a legitimate object for participation in the first place, what outcomes of a participation process will be made public and how they will be made public (Elam et al., 2007).

However, there are more expansive ways to interpret this metaphor which better account for the more disorganised and contingent realities of experimentation. Philosophers and historians of science have also drawn attention to the messy social processes around supposedly bounded and tightly controlled experiments. The historian of physics Peter Galison (1987) argued that the decision to end a particular experiment, when it is believed that enough evidence has been accumulated to authoritatively prove or disprove an assertion, is fundamentally a social one. In other words, even in the world of physics it is impossible to demonstrate that a question has been definitively closed or that disturbing effects have been completely removed from the experimental setting. Furthermore, in relation to the history of the natural sciences, the philosopher Ian Hacking (1983) characterises the relationship between experiments and scientific theories as a recursive one, with neither aspect entirely determining the other. Experiments may be interpreted and designed in line with certain working theories or more general concepts, but the outcomes of experiments may also lead to the creation of new conceptual objects which become involved in the design of new experiments. Consequently, in Hacking's account, scientific theory, experimental practice and broader social influences interact to produce scientific facts and to initiate new experiments, rather than one factor being the driving force.

This recognition of the never quite complete nature of experimentation has also been productively elaborated by geographers and STS scholars to draw attention to the continual existence of overflows from experimentation over space and time. Several authors have empirically demonstrated the ways in which scientific experiments go beyond the boundaries of the laboratory through the application of new technologies and knowledges to 'real-world' problems and their concurrent effects on modes of social organisation (Hinchliffe et al., 2005; Schwartz and Krohn, 2011; Szerszynski, 2005). This has led the geographer Gail Davies (2010) to argue that it is important to question not only when experiments end but also where they end. Others have labelled this as a broader regime of collective experimentation (Felt and Wynne, 2007) or as a series of 'wild experiments' (Lorimer and Driessen, 2014), evoking a picture of a multiplicity of open-ended experiments playing out at multiple scales within a democratic regime, with varying degrees of intentionality by the actors involved. In parallel to this, Callon, Lascoumbes and Barthe (2009) have elaborated the phenomena of 'participation in the wild', evoking a similar overflowing of processes and effects from the participation laboratory.

The concept of a regime of collective experimentation has particular resonance with discussions of governance or democratic innovations. As laid out in the European Commission report, 'Taking the European Knowledge Society Seriously' (Felt and Wynne, 2007), collective experimentation is characterised by the broad distribution of innovation across different actors and communities, trying out novel responses to societal challenges and learning from their repeated attempts or experiments both in innovation itself and in the governance of innovation. The term 'collective experimentation' has also been used by Bruno Latour (2011), but with a broader meaning encompassing the system-wide consequences of human interventions, such as climate change. Furthermore, Matthias Gross (2010a, 2010b) has built on this work to show the relevance of concepts of collective experimentation to ecological interventions on the ground, enacted by experts, citizens and policy actors in collaboration.

Democracy itself had been conceptualised as an experiment perhaps most influentially in John Dewey's (1927) book *The Public and Its Problems*, where he argues that even the formation of the state is an experimental process, consisting of trials and accidents around new rules and organisations. Publics come into being and become organised in response to new policy problems which are made up of these unforeseen indirect consequences, leading to new political institutions and forms of organisation. Crucially, the experiment must constantly be repeated as the state is continually discovered anew as novel challenges and publics arise. In this view democracy is not a stable entity or benchmark, but rather something that is continually practiced and contested (cf. Gallie, 1956). Understandings of democracy are not static but change over time in response to the emergence of new public problems and following developments in science and technology. Furthermore, there is change in collective understandings of who is being represented in a democracy, the appropriate relationship between the state and its citizens and what sort of knowledge can legitimately be drawn on in political processes (Jasanoff, 2011).

5.3 Experimenting with Democratic Innovations

It could be said that through experiments like climate governance and public participation new meanings of democracy, new definitions of public problems and new modes of engaging citizens are being produced. The concept of democratic innovations offers one analytic lens through which to describe and understand these processes, as well as drawing attention to the broader reconfigurations of relationships between citizens, science and the state which emerge from the multiple, open-ended and overlapping experiments of democratic governance in societies infused with science and technology (Callon et al., 2009; Ezrahi, 1990; Jasanoff, 2011).

Concepts like governance innovations and democratic innovations owe much to a broader literature which focuses on social innovations, broadly defined as innovations in social practices or forms of social organisation such as institutions (cf. Pol and Ville, 2009). The key conceptual move made here is to argue that particular procedures or bundles of practices can become more or less standardised – just like other forms of innovations – and can therefore travel into different political contexts and have multiple indirect effects as has been noted in the broader innovation literature (Fagerberg, 2006). Discussions of social innovation have been particularly significant in the fields of environmental governance, and citizen action and participation (e.g. Gottweis, Braun and PAGANINI, 2007; Seyfang and Smith, 2007). The concept does at least two important things with regards to understandings of governance. First, accounts of social innovations often draw attention to issues of justice and societal well-being, rather than purely technically defined societal aims and measures of progress (Dawson and Daniel, 2010). Secondly, and most significantly, the concept highlights that progress in addressing societal challenges such as climate change is not only dependent on technical innovations and fixes, but also will be achieved – or perhaps even primarily achieved – through changes in social practices and forms of organisation (Seyfang and Smith, 2007). Thus, the concepts of social, governance and democratic innovations are a useful challenge to dominant technically led or even technocratic visions of energy transitions and climate change governance.

The political scientist Graham Smith was one of the earliest and probably among the most prominent analysts to offer a definition and empirical elaboration of the term 'democratic innovations' in his book of the same name (2009). Smith defines democratic innovations as 'institutions that have been specifically designed to increase and deepen citizen participation in the political decision-making process (1), using the concept to describe and analyse standardised modes of citizen participation including participatory budgeting and citizens panels. For Smith these practices are innovative in that they represent a departure from conventional institutional architectures. However, in light of broader literatures on technical and social innovations there is more to unpack in this concept of democratic innovations.

First, there are other social innovations which could also be labelled as democratic innovations, beyond the invited deliberative processes of citizen participation which Smith considers. Established governmental modes of citizen engagement such as consultations and public opinion polls were once novel democratic innovations, whilst more emergent policy practices of citizen engagement such as open policy approaches and open data could be considered as democratic innovations in the making. Furthermore, whilst Smith states that his concept of democratic innovations concerns only highly institutionalised practices, it is also possible that democratic innovations could become standardised and institutionalised outside of formal governing institutions. For example, practices of community organisation such as co-operatives or community currencies, alongside widely practiced forms of activism such as direct action or petitioning, could become standardised and might also be considered to be democratic innovations (cf. Seyfang and Smith, 2007).

Secondly, there is potential mileage in taking the 'innovation' part of democratic innovations more seriously. Work in STS, innovation studies and the social studies of technology has offered further insights into processes of innovation by giving detailed histories of innovation processes which highlight their contingencies and the social processes which have shaped them (e.g. Pinch and Bijker, 1987), as well as pointing to their potential future trajectories and identifying path dependencies (Berkhout, 2002; Fagerberg, 2006). Furthermore, many accounts have highlighted the unintended consequences which emerge from technological innovations, sometimes with very significant impacts for society (Irwin, 1995), as well as the broader effects which innovation processes have on modes of social organisation (Jasanoff, 2004). This has led to many calls for and attempts to orchestrate procedures for what has been labelled as anticipatory governance, reflexive governance (Beck, 1994), constructive technology assessment (Schot and Rip, 1997), and – most recently – responsible innovation (Stilgoe, Owen and Macnaghten, 2013) in order to more fully take into account the social dimensions and potential effects of innovations. This work suggests that it might be possible as well to think of democratic innovations as also having these complex and contingent social histories, potential trajectories and path dependencies and broader societal effects. If this is the case, there is an important role for academic analysis in identifying and describing these dimensions, as well as finding ways to anticipate, govern and be responsible for the potential effects of these innovations.

A small group of STS scholars has already made some advances in this project, held together under the labels of technologies of participation or democracy. Using this conceptual framework, authors have examined the genealogies or contingent histories of prominent technologies of participation such as the focus group (Lezaun, 2007), as well as the kinds of participants (Lezaun and Soneryd, 2007; Marres and Lezaun, 2011), issue framings (Marres, 2007) and forms of action (Laurent, 2011) which these technologies produce. Taking further inspiration from innovation studies, Soneryd (2016) has explored how one such technology of participation travelled and was translated into a new national and political context, with implications for its broader social effects and attendant modes of social organisation. Furthermore, these scholars have also started to consider what implications their findings have for the governance of participation, with Voβ (2016) carrying out the first constructive assessment process concerned with participatory methods, and Chilvers and Kearnes (2016b) calling for a broader transformation in modes of participation towards a more reflective and reflexive approach. Whilst using the metaphor of a technology to describe procedures of participation arguably runs the risk – similar to the laboratory of participation metaphor – of overemphasising the bounded and controlled nature of most processes of participation, the situation of these studies within wider work on technological innovation which emphasises the constant production of unintended effects and new forms of social organisation also evokes the sense of open-ended experimentation described at the end of Section 5.2. Many of these authors have also increasingly adopted a language of experimentation (Chilvers and Kearnes, 2016a) to point to the wider experiments in participation and democracy of which these technologies are a part.

5.4 Public Dialogue as Democratic Innovation

5.4.1 Innovations in Participation

At the start of the twenty-first century many of the important institutional developments around science and science policy concerned the putative move towards initiating more of a two-way dialogue with citizens around scientific issues which affected their lives. In the United Kingdom this move was characterised as a shift from a focus on the public understanding of science (PUS) towards an emerging approach which emphasised public engagement with science (PES) (Michael, 2011; Pieczka and Escobar, 2013). During the 1990s PUS activities were subject to robust academic criticism, labelling their way of engaging with citizens as the 'deficit model' which assumed citizens were empty vessels needing to be filled with the correct information in order to accept scientific advances and policy (Irwin, 2001; Owens, 2000). In the year 2000, in the wake of several large public science controversies, such as the BSE (Bovine spongiform encephalopathy or Mad Cow disease) crisis and the MMR vaccine, and the apparent failure of the PUS project, the House of Lords Science and Technology Committee produced the report 'Science and Society' (House of Lords, 2000) which called for direct dialogue with the public to become an integral part of science policymaking. At the time, as well as in later accounts, this report was viewed as a pivotal moment in democratic practice around science policy in the United Kingdom, setting in motion the institutionalisation of a more dialogic form of public engagement with science and science policy (Bickerstaff et al., 2010; Miller, 2001). The House of Lords report stimulated discussion in the UK government calling for public dialogue around science policy (HM Treasury, 2004; POST, 2001, 2002). Perhaps most significantly, this call for direct public involvement in science policymaking resulted in the creation of Sciencewise as part of the Science and Innovation Investment Framework 2004–2014 (HM Treasury, 2004), within what was then the Department for Trade and Industry.

The narrative of the increasing turn to more deliberative and dialogic modes of public interaction in government science policy, in response to public knowledge controversies and distrust of government experts and policies, has been continually reinforced in government documents and in academic work (Pieczka and Escobar, 2013). However, this story arguably obscures the labour of academic and political advocates and the broader political context around the time of the mooted shift from PUS to PES. Social scientists played an important role in advocating and developing deliberative approaches to public participation. In the United Kingdom, the institutional move from PUS to PES was supported by establishment figures who had long been involved in supporting government PUS projects, such as the historian of science John Durant and prominent critics of the PUS approach, such as the STS scholar Brian Wynne, both of whom acted as witnesses for the 'Science and Society' report (House of Lords, 2000). STS scholar Charles Thorpe (2010; Thorpe and Gregory, 2010) has also argued that the turn towards public deliberation, and particularly the focus on reaching consensus as a key aim in processes like Sciencewise's public dialogues, is part of the broader development of the post-Fordist public in post-industrial British politics. Thorpe particularly highlights the role of the prominent sociologist Anthony Giddens and the left-wing think tank Demos in laying the groundwork and working with the New Labour project to bring consensus politics and participatory democracy centre stage (Thorpe, 2010).

As described in more detail in Pallett and Chilvers (2013), the Sciencewise programme has undergone several distinct phases in its existence, through which its approach to the practice of public dialogue has evolved as well as its relationship to UK government policymaking. In its early phase the programme supported projects which were often quite experimental in nature, creating card games or plays for youth engagement around science policy topics, and tended to lack clear connections to government policy decisions. After the programme's relaunch as an expert resource centre in 2006, the dialogue projects it supported became much more standardised, and a clear definition of public dialogue emerged featuring extended deliberation over one or more days; two-way dialogue between experts, policymakers and the public; workshops held in different parts of the country with a roughly demographically representative but small group of citizens; the creation of 'balanced' introductory materials for the participants to give them a good understanding of the issues under discussion; advisory structures overseeing the process and the materials the participants were presented with; and a clear policy hook with a government department, agency or research council which the dialogue outputs would feed into.

The practice of public dialogue which was developed took inspiration from what were considered successful examples of public engagement in other European countries, most significantly the Danish Board of Technology and the Rathenau Institute in the Netherlands. The method can also be seen as an at times uneasy amalgamation of earlier democratic innovations which were seen as legitimate in the UK government context, namely focus group methods and public opinion polling. Thus, whilst public dialogues are essential small group deliberative workshops, they are still required to be broadly demographically representative of the UK population – like a public opinion survey – even though the numbers involved are too small to be statistically significant.

The most recent (2012–2016) contract period of the Sciencewise programme saw further changes in institutional arrangements and the practice of public dialogue, as discussed in more detail in Pallett (2015). This contract period saw the involvement of the British Science Association and the participation-focussed body Involve in the day-to-day running of the programme, alongside the original contractor, the consultancy firm AEA. This, alongside the emergence of the open policy debate around the UK government during this time, led to further experimentation around the methods and institutional contexts of public dialogue processes and involved Sciencewise actors in broader debates about democratic governance and science policy in the United Kingdom.

5.4.2 Innovations in Climate Governance

Since its relaunch as an expert resource centre for public dialogue in 2008, Sciencewise was involved in a number of high-profile public dialogue projects around significant UK climate policy issues. The Big Energy Shift (2008–2009) was the first public dialogue supported and co-funded by Sciencewise and DECC. The dialogue reflected emerging policy interest in behaviour change at the time as a key element of UK climate policy and aimed to explore why people's behaviours were not changing in the expected ways in response to information provision and social marketing. This dialogue was important in establishing a productive long-term relationship between Sciencewise and DECC and was credited with spurring the creation of a further DECC-Sciencewise dialogue called the Low Carbon Communities Challenge (LCCC) (2010–2011).

In reality the LCCC was influenced by broader policy agendas concerning behaviour change and the emergence in the United Kingdom of community energy. However, the focus on engaging with already active community groups rather than individuals (as is usually the case in a public dialogue) was partly justified as a response to the key finding of the Big Energy Shift: that people's behaviours shift as part of groups rather than on an individual basis. Thinking on behaviour change in other parts of government was also starting to converge on the potential effectiveness of community-based initiatives, so policy actors saw an opportunity to harness the dynamism of community energy projects to promote behaviour change, as in the LCCC. However, this blurring between the behaviour change agenda and the community energy movement, which was characterised by a very different vision of the energy system and energy futures and focussed on energy supply, was also seen as problematic by some academic and civil society commentators. Ultimately, whilst the LCCC was one of the most high-profile public engagement projects around climate governance in the United Kingdom, the findings and impacts of the project were disrupted by the change of government in 2010, which signalled a change in tack on climate and energy policy.

At around the same time, Sciencewise and DECC supported another public dialogue project focussed on realising the 2050 greenhouse gas reduction targets laid out in the 2008 Climate Change Act. Here, the public dialogue was carried out alongside an expert process and economic assessment and was concerned with identifying potential pathways through which the targets could be reached and gauging their level of fit with public values. The dialogue's format also contained a number of novel elements including a separate youth panel which met several times during the process and the creation of an online game which allowed anyone who was interested to join in and explore the possible demand and supply-side options for reaching the targets. This game also structured the public dialogue workshops which focussed on the creation of group and, in some cases, individual-preferred pathways to reaching the 2050 targets using these different options.

After the start of the term of the Conservative-Liberal Democrat coalition government in the United Kingdom, the DECC-Sciencewise relationship continued around less high-profile projects, including some related to the Climate Change Committee and carbon budgets, which continued to address many of the concerns and ideas from the earlier 2050 pathways dialogue. In 2013 the Office of Unconventional Gas and Oil, which sat within DECC, also supported a public dialogue project related to the public controversy around fracking in the United Kingdom in partnership with Sciencewise. The dialogue focussed on the narrow issue of how to best compensate communities affected by fracking, rather than allowing a broader discussion of public concerns and values related to the issue.

Sciencewise has also partnered with many other parts of government to support public dialogue projects related to climate policy. For example, a number have been carried out with and for UK government research councils. A public dialogue project took place alongside the Living with Environmental Change cross-research council programme, exploring public responses to climate change research through a number of workshops in 2010. With the Biotechnology and Biological Sciences Research Council (BBSRC), Sciencewise supported the Bioenergy Distributed Dialogue (2012–2013) focussed around public responses to research into biofuels, a potential but at times controversial contributor to the low-carbon transition. This project also attempted to create new methodologies for public dialogues which would allow more sustained and long-term engagement with the public, as well as a more iterative relationship between the development of the BBSRC's research agenda and the outputs of dialogue workshops. This was achieved through a card deck which could be altered and added to in response to new research or public responses and could be downloaded and used by any group wanting to engage with research on biofuels.

As well as research councils, Sciencewise also partnered with government agencies to carry out public dialogue projects around climate policy. In collaboration with the Environment Agency and other bodies, Sciencewise supported a public dialogue on the topic of flood preparedness (2012–2013), focussing not only on the technical aspects of flood response but also on the need for institutional connections and responsiveness in dealing with flood risk and flooding events.

5.4.3 A Travelling Innovation

The previous account demonstrates how public dialogue as a democratic innovation has successfully travelled to and, in some cases, become embedded in different parts of UK government climate policy. Public dialogues have fed into climate change research, engaged with controversial topics and technologies like fracking and bioenergy and been used in relation to concrete pieces of legislation like the Climate Change Act, as well as contributing towards broader policy agendas, such as behaviour change. Beyond Sciencewise's direct interventions, public dialogue has also been taken up more broadly as a credible and effective method of public engagement by UK research councils, EU research programmes, market research companies, local government and even other governments, including the devolved Scottish government and the Japanese government. In these ways public dialogue as a democratic innovation has had broader impacts on climate change policy and governance, contributing towards decision-making around specific issues as well as processes of envisioning energy and climate futures.

Beyond individual processes of public dialogue, it is also possible to follow its wider effects on climate governance through widespread usage. While there are variations between processes, public dialogues tend to produce similar visions of citizens and their role in climate governance and similar framings of the climate 'problem', as well as resting on a particular vision of democratic engagement. The model of democracy which characterises this particular democratic innovation is deliberative and consensual, obscuring arguments and actions such as protests and public debates which are more antagonistic in nature – not uncommon around climate change. Furthermore, whatever the climate issue under discussion, the design of public dialogues processes – which makes a strong distinction between public and expert inputs – tends to present climate and energy as primarily technical issues. This is both influenced by and reinforcing dominant visions in climate governance of climate change being a technical issue that requires public behaviour change and acceptance of new technologies, rather than a broader exploration of public values and actions. Therefore, the publics of public dialogue projects are often imagined in fairly passive roles and are often seen as having little relevant knowledge of the issues prior to their involvement in the public dialogue process. This strong vision again has resonance and effects more generally in climate governance, limiting the potential roles citizens could play in the transition to a low-carbon and climate-adapted world. However, in these repeated experiments there is always room for contingency and emergence because the experiment is never quite finished. In some instances public dialogues have adopted different visions of the public or the issue in hand; for example, the LCCC was predicated on having much more active and knowledgeable participants. And in other instances, assumptions about publics and climate issues have been transformed and challenged during and after the process, for example, through unexpected public responses or the take up of public dialogue outputs in unexpected places, such as by civil society organisations.

As an outcome of repeated and overlapping experiments, the impacts of this democratic innovation cannot be limited to the realm of climate governance – even when only climate change-related public dialogues are considered. In the examples given previously, public dialogue processes also contributed to changing institutional attitudes towards public engagement and the public, for example, within DECC and the BBSRC. Through the constant repetition and travel of the public dialogue technique there has also been methodological innovation in order to adapt it to new contexts and aims, or as a result of learning. These new methods create the potential for new forms of engagement – such as the sustained and evolving engagement attempted in Bioenergy Distributed Dialogue – and therefore new ways for citizens to influence policy and governance.

5.5 Participation Overflowing the Laboratory: Going Beyond Experiments?

There are multiple forms of experimentation at play in this account of the institutionalisation of public dialogue and its effects on climate governance. The initial development of public dialogue as a democratic innovation can be described as an experimental process, requiring time and repetition to refine the procedure and approach during the early stages of the Sciencewise programme. Furthermore, this account has demonstrated that experimentation with the approach is ongoing, as it continues to shift to address new contexts, aims and challenges. In common with the existing literature this chapter also finds that there is analytical value in treating individual public dialogue processes as experiments. Moreover, they are experiments with a number of different concerns, including testing the credibility and acceptability of a set of policy propositions (testing hypotheses), trying out new working relationships (learning by doing), testing new techniques and tools (selecting designs that work) (cf. Turnheim et al., Chapter 1) – such as the 2050 pathways game and trying different ways to influence policy actors and processes. Public dialogue processes have also been part of broader experiments in climate governance, including the internationally unprecedented UK Climate Change Act, the community energy movement and government attempts to encourage the emergence of a UK fracking industry.

Far from being discrete and controlled, these experiments in participation and climate policy have overflowed the walls of their metaphorical laboratories in a number of ways. They have contributed to social movements, like community energy, beyond the bounds of formal climate policy or invited public engagement.

They have also helped shape and change institutions, for example, changing the BBSRC's attitude towards the value of public engagement in the case of the bioenergy distributed dialogue or showing related government agencies how they could work more effectively together in the case of the flood response dialogue. They have also had broader effects by contributing to the closing down of certain debates and issue spaces. For example, the fracking dialogue supported by Sciencewise and DECC arguably contributed towards a broader narrowing of the terms of debate about fracking, which denied the validity of broader discussions about public trust or social justice, or the role of fracking in the transition to a low-carbon energy system.

Examining experiments around public engagement and climate governance through the lens of one democratic innovation – public dialogue – also illuminates the role of experimentation in processes of organisational and policy learning. Not only do experiments generate new knowledge through testing hypotheses about the world but also they might contribute to learning by forging new ways of doing climate governance and public engagement, new relationships, new ways of organising and categorising knowledge and activities and new collective stories. The metaphor of experimentation also helpfully captures the non-linearity and ambiguous nature of these learning processes. Whilst it might have appeared in the context of some public dialogue projects and policy decisions that visions of the role of citizens in climate governance had shifted – for example, the LCCC's adoption of a model of active communities providing climate change solutions – other processes and decisions were an apparent backward step. Uncertainty and contingency are ever-present features of experimentation and innovation, and an understanding of the political dimensions of these experiments can often be the most useful element in identifying reasons for their relative success or failure, rather than their levels of embeddedness or standardisation.

The experiments described here are both open-ended and overlapping, influencing and connecting governance at multiple scales, including states, transnational bodies, local governance and individual institutions. This account also demonstrates the overlapping of issue areas through these experiments, including climate, energy, bioscience and risk and vulnerability. Even where there are no wider effects of a particular experiment yet in evidence, there is always potential for impact and influence to occur long after the initiation of the experiment. For example, elements of particular public dialogue processes get recycled and reused, as is the case with the pathways game created for the 2050 pathways dialogue. Furthermore, the outputs of different experiments which may have appeared irrelevant, inconvenient or useless at the time may be interpreted differently when taken up within a different institution or movement, or in the light of subsequent unexpected events.

It should not be a surprise that in the context of public engagement and climate governance, the experiment is never finished. Rather, constant monitoring and intervention are needed for the experiment to be reshaped and retried.

References

Beck, U. (1994). The reinvention of politics: Towards a Theory of reflexive modernization. In U. Beck, A. Giddens and S. Lash, *Reflexive Modernization Politics Tradition and Aesthetics in the Modern Social Order*. 1–55. Cambridge: Polity Press.

Berkhout, F. (2002). Technological regimes, path dependency and the environment. *Global Environmental Change*, 12, 1–4.

Bickerstaff, K., Lorenzoni, I., Jones, M., and Pidgeon, N. (2010). Locating scientific citizenship: The institutional contexts and cultures of public engagement. *Science Technology & Human Values*, 35(4), 474–500.

Bogner, A. (2012). The paradox of participation experiments. *Science, Technology & Human Values*, 37(5), 506–527.

Brown, M. B. (2009). *Science in Democracy: Expertise, Institutions and Representation*. Cambridge, MA: MIT Press.

Callon, M., Lascoumbes, P., and Barthe, Y. (2009). *Acting in an Uncertain World: An Essay on Technical Democracy*. Translated by G. Burchell. Cambridge, MA: MIT Press.

Chilvers, J. (2008a). Deliberating competence: Theoretical and practitioner perspectives on effective participatory appraisal practice. *Science, Technology & Human Values*, 33(2), 155–185.

Chilvers, J. (2008b). Environmental risk, uncertainty, and participation: Mapping an emergent epistemic community. *Environment and Planning A*, 40(12), 2990–3008.

Chilvers, J., and Kearnes, M. (2016a). Participation in the making: Rethinking public engagement in co-productionist terms. In Chilvers, J., and Kearnes, M. (eds.), *Remaking Participation: Science, Environment and Emergent Publics*. London: Routledge, 31–63.

Chilvers, J., and Kearnes, M. (2016b). Remaking participation: Towards reflexive engagement. In Chilvers, J., and Kearnes, M. (eds.), *Remaking Participation: Science, Environment and Emergent Publics*. London: Routledge, 261–288.

Cooke, B., and Kothari, U. (2001). *Participation: The New Tyranny?* London: Zed Books.

Davies, G. (2010). Where do experiments end? *Geoforum*, 41(5), 667–670.

Dawson, P., and Daniel, L. (2010). Understanding social innovation: A provisional framework. *International Journal of Technology Management*, 51(1), 9.

Dewey, J. (1927). *The Public and Its Problems*, New York: H. Holt.

Elam, M., Reynolds, L., Soneryd, L., Sundqvist, G., and Szerszynski, B. (2007). *Mediators of Issues and Mediators of Process*. Brussels.

Ezrahi, Y. (1990). *The Descent of Icarus*. Cambridge, MA: Harvard University Press.

Fagerberg, J. (2006). Innovation: A guide to the literature. In Fagerberg, J., Mowery, D. C., and Nelson, R. R. (eds.), *Oxford Handbook of Innovation*. Oxford: Oxford University Press, 1–26.

Felt, U., and Wynne, B. (2007). Taking the European Knowledge Society Seriously. *Report of the Expert Group on Science and Governance to the Science, Economy and Society Directorate, Directorate-General for Research*. Brussels: European Commission, 96.

Galison, P. (1987). *How Experiments End*. Chicago: University of Chicago Press.

Gallie, W. B. (1956). Essentially contested concepts. *Proceedings of the Aristoliean Society*, 56, 167–198.

Gisler, P., and Schicktanz, S. (2009). Introduction: Ironists, reformers, or rebels? Reflections on the role of the social sciences in the process of science policy making. *Science, Technology and Innovation Studies*, 5, 5–17.

Gottweis, H., Braun, K., and PAGANINI Consortium (2007). *Participatory Governance and Insitutional Innovation*. Final Report. Accessed 25/10/2017, www.univie.ac.at/LSG/paganini/finals_pdf/WP8_FinalReport.pdf

Gross, M. (2010a). *Ignorance and Surprise: Science, Society and Ecological Design*, Cambridge, MA: MIT Press.

Gross, M. (2010b). The public proceduralization of contingency: Bruno Latour and the formation of collective experiments. *Social Epistemology*, 24(1), 63–74.

Hacking, I. (1983). *Representing and intervening: Introductory topics in the philosophy of natural science*. Cambridge: Cambridge University Press.

HM Treasury (2004). *Science and Innovation Investment Framework: 2004–2014*. Norwich: The Stationary Office.

Hinchliffe, S., Kearnes, M., Degen, M., and Whatmore, S. (2005). Urban wild things: A cosmopolitical experiment. *Environment and Planning D: Society and Space*, 23(5), 643–658.

House of Lords (2000). Science and Society: The House of Lords Science and Technology Committee. Third Report.

Irwin, A. (1995). *Citizen Science: A Study of People, Expertise and Sustainable Development*. London: Routledge.

Irwin, A. (2001). Constructing the scientific citizen: Science and democracy in the biosciences. *Public Understanding of Science*, 10(1), 1–18.

Jasanoff, S. (2004). The idiom of co-production. In *States of Knowledge: The Co-production of Science and Social Order*. S. Jasanoff, ed. Abingdon, Oxon: Routledge, 1–12.

Jasanoff, S. (2011). Constitutional moments in governing science and technology. *Science and Engineering Ethics*, 17(4), 621–638.

Latour, B. (2011). From multiculturalism to multinaturalism: What rules of method for the new socio-scientific experiments. *Nature and Culture*, 6(1), 1–17.

Laurent, B. (2011). Technologies of democracy: Experiments and demonstrations. *Science and Engineering Ethics*, 17(4), 649–666.

Lezaun, J. (2007). A market of opinions: The political epistemology of focus groups. *Sociological Review*, 55, 130–151.

Lezaun, J., and Soneryd, L. (2007). Consulting citizens: Technologies of elicitation and the mobility of publics. *Public Understanding of Science*, 16(3), 279–297.

Lorimer, J., and Driessen, C. (2014). Wild experiments at the Oostvaardersplassen: Rethinking environmentalism in the Anthropocene. *Transactions of the Institute of British Geographers*, 39, 169-181

Marres, N. (2007). The issues deserve more credit: Pragmatist contributions to the study of public involvement in controversy. *Social Studies of Science*, 37(5), 759–780.

Marres, N., and Lezaun, J. (2011). Materials and devices of the public: An introduction. *Economy and Society*, 40(4), 489–509.

Michael, M. (2011). 'What are we busy doing?': Engaging the Idiot. *Science, Technology & Human Values*, 37(5), 528–554.

Miller, S. (2001). Public understanding of science at the crossroads. *Public Understanding of Science*, 10, 115–120.

Munton, R. (2003). Deliberative democracy and environmental decision-making. In Berkhout, F., Leach, M., and Scoones, I. (eds.), *Negotiating Environmental Change: New Perspectives from Social Science.* Cheltenham, UK: Edward, Elgar, 109–136.

Osborne, T. (2004). On mediators: Intellectuals and the ideas trade in the knowledge society. *Economy and Society*, 33(4), 430–447.

Owens, S. (2000). 'Engaging the public': Information and deliberation in environmental policy. *Environment and Planning A*, 32(7), 1141–1148.

Pallett, H. (2015). Public participation organizations and open policy: A constitutional moment for British democracy? *Science Communication*, 37(6), 769–794.

Pallett, H., and Chilvers, J. (2013). A decade of learning about publics, participation and climate change: Institutionalising reflexivity? *Environment and Planning A*, 45(5), 1162–1183.

Parliamentary Office of Science and Technology (POST) (2001). *Open Channels: Public dialogue in Science and Technology.* London: POST. www.parliament.uk/documents/post/pr153.pdf

Parliamentary Office of Science and Technology (POST) (2002). *Public Dialogue On Science & Technology*, London: POST. Accessed, 25 October 2017, www.parliament.uk/documents/post/pn189.pdf

Peck, J., and Theodore, N. (2015). *Fast Policy: Experimental Statecraft at the Thresholds of Neoliberalism.* Minneapolis: University of Minnesota Press.

Pieczka, M., and Escobar, O. (2013). Dialogue and science: Innovation in policy-making and the discourse of public engagement in the UK. *Science and Public Policy*, 40(1), 113–126.

Pinch, T. J., and Bijker, W. E. (1987). The social construction of facts and artifacts: Or how the sociology of science and the sociology of technology might benefit each other. In Bijker, W. E., Hughes, T. P., and Pinch, T. J. (eds.), *The Social Construction of Technological Systems: New Directions in the Sociology and History of Technology.* Cambridge, MA: MIT Press, 17–50.

Pol, E., and Ville, S. (2009). Social innovation: Buzz word or enduring term? *Journal of Socio-Economics*, 38(6), 878–885.

Schot, J., and Rip, A. (1997). The past and future of constructive technology assessment. *Technology Forecasting and Social Change*, 54(1996), 251–268.

Schwartz, A., and Krohn, W. (2011). Experimenting with the concept of experiment: Probing the epochal break. In Nordmann, A., Radder, H., and Schiemann, G. (eds.), *Science Transformed? Debating Claims of an Epochal Break.* Pittsburgh, PA: University of Pittsburgh Press, 119–134.

Seyfang, G., and Smith, A. (2007). Grassroots innovations for sustainable development: Towards a new research and policy agenda. *Environmental Politics*, 16, 584–603.

Smith, G. (2009). *Democratic Innovations: Designing Institutions for Citizen Participation.* Cambridge: Cambridge University Press.

Soneryd, L. (2016). Technologies of participation and the making of technologised futures. In Chilvers, J., and Kearnes, M. (eds.), *Remaking Participation: Science, Environment and Emergent Publics.* London: Routledge, 144–161.

Stilgoe, J., Owen, R., and Macnaghten, P. (2013). Developing a framework for responsible innovation. *Research Policy*, 42(9), 1568–1580.

Szerszynski, B. (2005). Beating the Unbound: Political Theatre in the Laboratory Without Walls. In Giannachi, G., and Stewart, N. (eds.), *Performing Nature: Explorations in Ecology and the Arts.* Frankfurt: Peter Lang, 181–197.

Thorpe, C. (2010). Participation as post-Fordist politics: Demos, new labour, and science policy. *Minerva*, 48(4), 389–411.

Thorpe, C., and Gregory, J. (2010). Producing the post-Fordist public: The political economy of public engagement with science. *Science as Culture*, 19(3), 273–301.

Voβ, J.-P. (2016). Reflexively engaging with technologies of participation: Constructive assessment for public participation methods. In Chilvers, J., and Kearnes, M. (eds.), *Remaking Participation: Science, Environment and Emergent Publics*. London: Routledge, 238–260.

6

Broadening Experimentation through Research-Industry Collaboratives in the Australian Water Sector

MEGAN A. FARRELLY AND JOANNETTE J. BOS

6.1 Introduction

The pace of change and growth in cities worldwide has led to urban environments becoming critical sites for examining how alternative technologies and/or practices can address increasing climate stresses. This is evidenced by a growing number of 'living laboratories', 'urban laboratories' and other city-focused research-industry collaborations that are experimenting with alternative technical and governance configurations to advance sustainability transitions (e.g. Evans and Karvonen, 2014; Nevens et al., 2014). Until recently, urban environments in predominantly developed cities had derived significant economic and societal benefits from large-scale, centralised, technical infrastructure, such as energy grids and urban water systems (i.e. supply, sanitation and drainage networks). However, these infrastructure systems are now regarded as ill-equipped to meet contemporary and future challenges posed by growing populations, increasing demand and global climate change predictions (see, e.g., Milly et al., 2008; Brown, Ashley and Farrelly, 2011). This has led to a groundswell of urban-based experimentation whereby innovative technologies and/or practices have been applied in an effort to avoid future maladaptations (see Barnett and O'Neil, 2010). Contemporary examples within Australia are found in two large-scale Cooperative Research Centres[1] (CRC) which focus on generating low-carbon-built urban environments (Low Carbon Living CRC) and ways to revolutionise urban water management (CRC for Water Sensitive Cities). These and other research-industry collaboratives are explicitly designed to bring stakeholders from research, the private sector and government authorities together with the aim of co-producing new knowledge, providing platforms for building collective

---

[1] Cooperative Research Centres are an Australian government-supported initiative, whereby industry, government and research organisations come together to explore new options for advancing and promoting innovation across a variety of different sectors (i.e. water, mining, health, agriculture, etc.).

understandings and generating action for change (Boardman and Gray, 2010). For example, the CRC for Water Sensitive Cities mission is to bring together researchers from universities across Australia and internationally to work along-side government and water-industry practitioners to experiment, synthesise new knowledge and influence practice by empowering a change in practice (see, e.g., https://watersensitivecities.org.au).

Research-industry collaboratives are typically at the frontier of knowledge production and developing/experimenting with alternative technologies and practices and are explicitly designed to promote technology and knowledge transfer to promote transformational change (Broadman and Gray, 2010). However, there has been limited explicit attention directed towards understanding the mechanisms and/or processes employed by these research-industry collaboratives to deepen (learning as much as possible about problem and solutions), broaden (where an approach is replicated in other contexts) and embed experimental insights (new ideas and approaches are sustained) within governance structures for responding to urban challenges such as climate change (see Bos and Brown, 2012; Turnheim, Kivimaa and Berkhout, Chapter 1). Therefore, this chapter reflects on two decades of collaborative experimentation within the Australian urban water sector and (1) attempts to characterise shifts in the 'type' of experiment undertaken over time and (2) unpacks the processes and strategies that led urban stormwater practices to operate a shift beyond experimentation and towards mainstream practices. First we briefly describe the (re)emergence of research-industry collaborative programmes to tackle broad urban sustainability issues and, in doing so, showcase how the research-industry collaborative is itself an experiment which brings disparate stakeholders and researchers together, all the while fostering and activating experimentation. Next, we reflect on the experiences of the Australian urban water sector to reveal how collaborative experimentation styles have (co)evolved, expanded and responded to political and climatic variability (i.e. the Millennium drought). We then turn to the overarching insights derived from these research-collaborative experiences by identifying a number of common processes and mechanisms used to support replication and assist with reframing approaches to embed new understandings and generate a shift away from conventional urban water governance. These reflections are drawn from the experiences of both authors, who have been engaged in urban water governance research, working closely with industry practitioners, for more than ten years. In addition, both authors have specifically examined aspects of experimentation in the urban water sector and, more recently, have been involved in undertaking evaluations of research-industry collaboratives, which has provided unique insights into the design and function of these platforms.

6.2 Research and Development Collaboratives and Urban Sustainability Experimentation

Advancing sustainable urban transformations demands a shift away from traditional linear innovation transfer models towards more complex, transdisciplinary collaborative initiatives that go beyond disciplinary boundaries to engage a variety of actors and/or organisations. Such collaborations have been referred to as 'post-normal science' (e.g. Funtowicz and Ravetz, 1993), mode-2 science (e.g. Gibbons et al., 1994), triple helix paradigms and systems (e.g. Etzkowitz and Leydesdorff, 2000 and Ranga and Etzkowitz, 2013, respectively). Collectively, these approaches recognise the need for alternative forms of knowledge production that engage with anticipated end users in an attempt to generate sustainable solutions and assist with their implementation (Lang et al., 2012; Wiek et al., 2014). Research-industry collaboratives have long played an important role in providing a 'safe space' for fostering experimentation and supporting innovation diffusion (e.g. Rip, 1995). In more recent times, these research-industry collaborations have been recognised for their critical role in advancing our understanding of sustainability sciences (Wiek et al., 2014; Dentoni and Bitzer, 2015), whereby these collaborative initiatives often function across multiple scales and link different areas of expertise and societal activities to inform change (see, e.g., Olsson, Folke and Hughes, 2008; Trencher, Yarime and Kharrazi, 2013; Trencher et al., 2014). In doing so, research-industry collaboratives appear to have multiple purposes with regard to experimentation: they foster technological and governance innovations but also experiment with different (and often multiple) forms of engagement mechanisms, reflecting the diversity of stakeholders involved.

Within this context, there has been a rise in the number of research and development programmes which directly incorporate industry and government involvement. These non-academic stakeholders are no longer just financial partners or receivers of scientific wisdoms, but rather they are enrolled as actors who assist with new knowledge creation, test knowledge in real-life settings and actively champion change efforts both within their organisation and more broadly. While the production of scientific knowledge is increasingly regarded as a 'transformational agent' (Perry, 2006: 202), and necessary in the age of 'evidence-based decision-making' (Head, 2008), the authors suggest that without the support of a collaborative (and coordinative) platform to engage with, share and discuss these scientific insights, their transformational power remains limited (see also Hölscher, Frantzeskaki and Loorbach, Chapter 7). Thus, for the purpose of this chapter, we consider the roles of research-industry collaboratives to be analogous to a 'bridging organisation' (Kampelmann, Van Hollebeke and

Vandergert, 2016), or an intermediary (Kivimaa, 2014). We argue that research-industry collaboratives operate as a formal network within a niche, with explicit ties to the regime, thus creating opportunities for technological and institutional innovation by bringing normatively aligned actors from different domains (research, planning, policy, operations, private enterprise), scales and decision-making levels (i.e. local and state governments) together to co-create, co-test and share insights regarding alternative practices and technologies (Folke et al., 2005; Crona and Parker, 2012). This infers that relationship building, rather than generic links or partnerships between/among science and industry stakeholders, plays a strong role in going beyond experiments by assisting with generating, translating and embedding innovations. Therefore, drawing on the different conceptualisations and interpretations of experimentation presented in Chapter 1, alongside the notions of deepening, broadening and scaling up (cf. Grin et al., 2010), this chapter showcases how, over time, research-industry collaboratives applied different experimentation approaches and fostered the deepening and broadening of experimentation practice which resulted in policy and practice-based change (1).

6.3 Transforming Urban Water Governance through Experimentation

Stormwater in Australia, as in many parts of the world, has been traditionally viewed as a nuisance and a flood risk, resulting in extensive underground drainage networks and channelised streams that efficiently convey stormwater to specific discharge sites (typically local water bodies such as creeks, lakes or bays). This approach was designed to provide flood protection for households, businesses and infrastructure, at the expense of natural waterways. However, since the 1980s there has been mounting evidence of the negative economic, social and environmental impacts resulting from these practices (e.g. Fletcher and Deletic, 2006). As a result, Australian urban water management has witnessed a transformation in the way stormwater as conceived, valued and governed (Bos and Brown, 2012; Brown, Farrelly and Loorbach, 2013; Ferguson, Brown and Deletic, 2013).

In the early 1990s, two separate, but highly interrelated research-industry collaboratives formed[2] to investigate issues surrounding challenges associated with degradation of urban waterways and ecosystem health. These collaboratives established close working relationships between industry partners and researchers from across eastern Australia (i.e. Canberra, New South Wales, Queensland and Victoria). Collectively, the collaboratives undertook traditional scientific,

---

[2] The Cooperative Research Centre for Catchment Hydrology (1992–2005) and the Cooperative Research Centre for Freshwater Ecology (1993–2005).

Table 6.1. *Different 'types' of experimentation within research-industry collaboratives*

Type of experiment		Description	Actors involved	Examples
Traditional linear R&D innovation approaches	Hypothesis testing	Traditional scientific approach to exploring a problem	Science leads and industry receives knowledge	- Studying the levels of nitrogen exportation from catchment - Determining ecological impact of elevated nitrogen levels
	Design Testing	Showcasing how alternative technologies and/or processes can address a problem, while assessing feasibility in shielded, 'safe spaces'	Science leads and works with key industry partners to 'pilot' the new technology/ approach	- Using wetlands as natural biofilters to adsorb excess nitrogen - Providing key demonstration projects (i.e. Lynbrook Treatment Train)
Transformational approach to going beyond experimentation	Learning-by-doing	Recognises the contextual challenges related to disrupting status quo, and the need for the broadening and embedding of experiments Interactive forums where actors share tacit technological, operational and governance-related experiences based on their organisational contexts	Science, industry and government actors work together	- Facilitating synthesis workshops, formal platforms for actors involved in new land developments to openly discuss potential options and pathways for embedding new practices

Source: Adapted from Turnheim et al. (Chapter 1)

'hypothesis-testing' experimentation, identifying key non-point source pollution loads and strategic prevention technologies for mitigating degraded urban waterways and ecosystem health. This resulted in credible scientific evidence that was reproduced in industry-relevant reports and guidelines (applicable to broad audiences) related to improved protection and management techniques (i.e. gross pollutant traps), strategies for waterway health, and 'whole-of-catchment scale' management. Lead researchers came to recognise their scientific evidence was an important source of policy advice, and through the networks and relationships forged during this period, industry and research actors alike were encouraged to become 'politically active' (i.e. media interviews and active lobbying for a change in practice). This required lead actors to tailor their scientific narrative to respond to the dominant political agendas of the time; shifting their emphasis from freshwater quality towards broader ecosystem health, and eventually to presenting stormwater as a legitimate resource to mitigate drought-impacts associated with climate change.

Building on the body of scientific knowledge generated and early stakeholder engagement successes, experimentation continued to simultaneously be nurtured within the protected boundary of the collaborative. This involved funding the piloting of innovative designs and technologies (i.e. gross pollutant traps, wetland filtration systems) alongside trialling collaborative forums for knowledge exchange (see Table 6.1). This required enrolling and working with actors beyond those directly involved in the collaboratives to secure local sites for in situ testing and showcasing new technological and management approaches for developers and local municipalities. This included design-testing a large-scale (regional) urban stormwater treatment wetland and a smaller, street-scape treatment version with the aim of understanding the associated practical design and construction challenges (see, e.g., Lloyd, Wong and Porter, 2002). The large-scale site was secured through support by an enrolled stakeholder organisation, which was actively promoting this alternative approach. However, securing an appropriate site for the smaller street-scape treatment train required extensive cooperation among a number of different stakeholders (many external to the collaborative). The local municipality was originally reluctant to accept any alternative technological approach to stormwater management for which they would have long-term responsibility. Negotiations ultimately led to a state government organisation agreeing to carry the risk should the experimentation fail (i.e. creating the safe space) (see Farrelly and Brown, 2011). This demonstration site, along with a traditional stormwater management site in neighbouring streets, was closely monitored and ultimately proved successful in detaining and carrying high volume stormwater flows, while also improving water quality and improving amenity values (see Lloyd et al., 2002).

Four important features were associated with extending beyond the 'design testing' experimentation processes, towards embracing a 'learning-by-doing' approach.

1. Dedicated monitoring and evaluating activities analysed experimental outcomes through different lenses (i.e. local economic impact on house prices, quality of the water and maintenance aspects; see Lloyd et al. [2002] for further information). This yielded important credible evidence and practical information to support replication and for shaping alternative narratives regarding the benefits of such a system.
2. Drawing on the collective scientific data, researchers worked alongside industry representatives to develop decision-support tools to assist with broader replications and adoption of these alternative stormwater technologies.
3. Researchers and industry stakeholders built on their nascent relationships and connections to the external policy environment to successfully lobby for and reshape the urban water policy landscape (e.g. the state-led, Victorian Stormwater Management Program launched in 2000 required municipalities to develop and implement stormwater improvement plans) (see Brown et al., 2013).
4. Formation of a dedicated capacity-building program (see www.clearwater.asn.au), which, in association with lead researchers and industry stakeholders from the collaboratives, ran facilitated 'site-visits' for intra- and interstate visitors. This contributed to a growing level of confidence among urban stormwater decision-makers, primarily local municipality actors, who recognised value in trialling this alternative approach. This capacity-building program continues to operate and has been replicated in other Australian capital cities. A key role of the capacity-building program is to (a) help showcase evidence to support adoption and (b) assist in (re)shaping the narrative around advancement of innovative urban water practices.

What occurred next was a growth in 'design-testing' replication – a broadening of the experiments (see Figure 6.1) – whereby, an increasing number of innovation champions (i.e. front runners) within local municipalities (see, e.g., Taylor et al., 2011) began experimenting within their own contexts, with the aim of generating localised experiences, understandings and assessments regarding whether this new approach fulfilled their own (and their organisations') objectives. The broadening of municipal level 'design-testing' experimentation was fuelled by (1) local municipal innovation champions who built internal networks to advocate within their organisations for experimenting with the new approach; (2) the launch of key state policy directives that brought attention to the need to improve urban stormwater management (i.e. Victorian Stormwater Management Program [2000–2007] and the New South Wales's Urban Stormwater Program [1998–2001]), and (3) land developers who began incorporating alternative stormwater management approaches into their on-ground works. The multiple experiments generated new, varied and reflective insights into what was feasible and possible within contemporary understandings.

In 2005, the previous collaboratives concluded. Nevertheless, many of the professional relationships and formal/informal networks formed over the life of these programs persisted (many still do today). Similarly, the capacity-building program provided ongoing opportunities to engage with 'learning-by-doing' experimentation by showcasing the growing number of successful and failed examples of alternative stormwater management systems. Furthermore, building on the foundational work of the preceding collaboratives, another two separate, but interrelated collaboratives formed: the technology-focused Facility for Advancing Water Biofiltration (FAWB) and the National Urban Water Governance Program (NUWGP). Of note, many industry and government stakeholders who had invested in the previous stormwater research also invested in both these research programs, alongside new interstate actors. FAWB focused on continuous design-testing and in situ learning-by-doing experimentation related to advancing stormwater treatment technologies. Similar to the preceding research programs, this resulted in the launch of industry guidelines for constructing and implementing the technology. Meanwhile, the NUWGP had identified significant challenges with regards to ongoing institutional path dependencies (see, e.g., Brown and Farrelly, 2009; Brown et al., 2011). Responding to these challenges, the NUWGP sought to isolate appropriate institutional mechanisms and governance configurations to underpin the delivery of more sustainable urban water management practices. These two programmes of work also continued to broaden their actor networks away from traditional water utilities and to involve land development companies, urban planners and aligned consultancy firms.

Responding to persistent drought conditions and a significant change in the political landscape (i.e. new leadership), leading Melbourne-based researchers and industry stakeholders (who had now built solid professional relationships and continued to operate within an informal network) dovetailed their research agenda, resources and on-ground activities to create an interdisciplinary Centre for Water Sensitive Cities (Wong and Brown, 2009; Brown, Wong and Deletic, 2015). This centre of research and learning engaged social scientists and engineers, but broadened to include climate scientists, landscape architects, economists, urban planners and medical scientists, among others. The broadening of research disciplines and industry partners reflects the connective capacity water provides within an urban environment (see Bettini et al., 2013). This was also an explicit attempt to maintain and broaden the relevancy of examining alternative urban water management pathways in a narrowing political-industry narrative towards water supply security due to prolonged drought conditions. This new research-industry alignment incorporated alternative ways of framing and conceiving stormwater. Within these programmes of research, stormwater was perceived as a potential resource, rather than simply a nuisance (i.e. flood risk), and 'design-testing' experiments extended beyond water quality issues to explore the potential of stormwater for keeping cities green and cool (microclimates); improved aesthetics and amenity; economic advantages; flood minimisation and increased recreational opportunities, among others.

Overall, the emerging narrative generated new conversations and helped form new industry, policy and political connections across a number of Australian states. The broadening of the stormwater narrative seeded great change, whereby the Centre for Water Sensitive Cities was quickly subsumed by an even larger collaborative (the CRC for Water Sensitive Cities 2012–2021). With the aim of revolutionising urban water management in response to climate variability, climate change and growing urbanisation pressures, the CRC for Water Sensitive Cities secured financial and in-kind support from more than eighty stakeholder organisations from international (China, Singapore and Netherlands), national, state and local government levels, alongside relevant industry stakeholders (multinationals to small-to-medium enterprises). Here, the interdisciplinary research agenda also expanded beyond stormwater to incorporate other potential urban water sources (i.e. recycled wastewater), all framed within the larger narrative of working towards delivering 'water sensitive cities' (see Brown, Keath and Wong, 2009; Wong and Brown, 2009). The research programme simultaneously investigates (design-testing) advances in technologies, while examining strategies for addressing societal, economic, institutional, planning, regulatory and policy opportunities and constraints related to delivering more sustainable, resilient, productive and liveable urban spaces (CRCWSC, 2013).

Alongside the design-testing experimentation, the programme design of the CRC for Water Sensitive Cities also built on lessons learned regarding stakeholder engagement mechanisms that arose from the preceding collaboratives. Avoiding traditional linear knowledge exchange, the CRC for Water Sensitive Cities focuses on (where possible) co-creating new empirical insights related to advancing sustainable water management through design-testing *and* learning-by-doing experimentation and (1) has invested in establishing and implementing a dedicated formative evaluation and learning framework. In addition, the program also (2) actively works to support and create adoption pathways for new knowledge and innovations through hosting and facilitating multiple, varied and highly interactive knowledge exchange platforms (e.g. national industry partner workshops, research synthesis activities and local/regional seminars, among others); (3) regularly engages with the political domain by providing timely policy advice, adapting narratives when/if required and lobbying/advocacy work; and (4) designs and delivers contemporary education programs based on the research insights for professional development and higher education.

6.4 Insights Across the Changing Research-Industry Collaboratives

What emerges over the course of this timeline is a broadening away from purely traditional hypothesis testing scientific experiments undertaken solely by researchers, funded by industry with unidirectional information flows, towards design-testing and experimenting with collaborative research approaches, whereby researchers and industry partners work together to craft research agendas, undertake experiments and share tacit knowledge to assist with embedding new knowledge into practice-based change. Over time, the collaboratives have played an important bridging role in facilitating and implementing various 'types' of experimentation and stakeholder engagement, which has helped embed innovative, alternative stormwater management technologies within policy and practice (see Brown et al., 2013). By employing strategic, tactical, operational and reflexive activities reminiscent of transition management (see Figure 6.1; see also Loorbach, 2010, the collaboratives generated fundamental change within the perspectives, structures and practices of the technological and governance regime (e.g. Bos, Brown and Farrelly, 2013; Brown et al., 2013).

Throughout a number of years, these efforts were repeated in increasingly more collaborative and sophisticated programs of research, culminating in the contemporary research-industry collaborative approach, where almost equal attention is directed towards (1) identifying and generating pathways for socio-political and institutional reforms; (2) developing new technological and methodological techniques for underpinning sustainable urban water management approaches; (3) developing guided experimentation and problem-solving projects; and (4) informing and engaging with broader institutional and technical learning and capacity-building programs. The multifunctional capacity of this collaborative space has proven to be a valuable model for broadening and embedding the reach of new knowledge and practices related to sustainable urban stormwater management and beyond.

Notwithstanding certain ongoing challenges, a number of key features have influenced the shift in the design and activity of collaboratives. Here, we consider three salient aspects for the conception and design of future collaboratives that can support a shift beyond experimentation.

6.4.1 Actor Networks

The emergent and persistent actor networks were open and responsive to, but also shaped, the growing sophistication regarding the way urban stormwater was

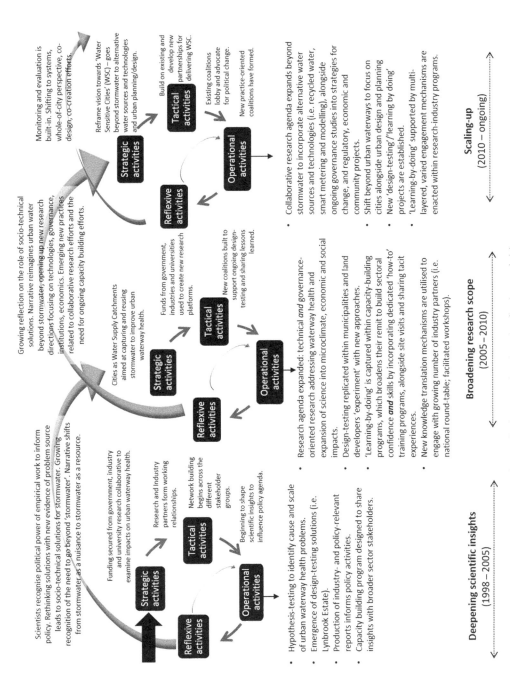

Figure 6.1. Stepwise incorporate strategic, tactical, operational and reflexive activities (adapted from Bos and Brown, 2012).

113

understood (i.e. as a potential resource). The collaborative platforms operated as a strategic, coordinative node that brought together a broad range of traditionally disparate actors to experiment with new ways of conceptualising and undertaking water management. The collaborative hubs can be considered as forms of governance innovation in themselves, insofar as they involve recasting the contours of the science-policy interface. Within these platforms, actors who would otherwise not normally interact were exposed to and engaged in discussion around emerging technological and governance innovations for advancing urban water practices and sharing experiences about what has and hasn't worked within their organisations and from the work-domain (i.e. planning, operations, asset management, etc.). The various collaborative hubs helped broker, build and support relationships through strategic system-building activities beyond the individual experiments, which in turn seeded other important formal and informal networks enabling further activities (e.g. Bos and Farrelly, 2015). By operating as a bridging organisation, the collaboratives brokered new network connections with other key coalitions: policy, capacity building, politics and technical, which collectively supported the embedding of new technologies and a practice change in urban water management (see, e.g., Bos, Brown and Farrelly, 2015).

6.4.2 Building and Maintaining Relationships

Much of the collaboratives' success with experimentation was initiated and sustained due to long historical professional relationships among a loose, informal network of boundary spanners working across science and industry (see e.g. Brown et al., 2013), which assisted in orchestrating the networks (e.g. Williams, 2010; Hölscher et al., Chapter 7). The creativity and responsiveness of many of the lead actors (research and industry) ensured that, despite political perspectives shifting, strong advocacy and lobbying activities were undertaken to (re)frame the narrative to suit the agenda of the time. Here, the policy entrepreneurs (c.f. Huitema and Meijerink, 2009) collectively contested the status quo through (1) generating relevant and salient evidence to underpin the arguments for creating change and (2) lobbying policy networks from the inside to stimulate change (the capacity to undertake this improved over time as individuals began to hold more senior decision-making positions). Their success lay in building a variety of relational capacities (i.e. trust and confidence) among and between different actors/organisations and across scales, which increased opportunities for joint action and was important for facilitating and supporting effective knowledge exchange and translation (i.e. deepening and broadening).

While relational capacity at the interpersonal level is critical, so too are the relationships between organisations. Often the assumption was made that individuals/organisations enrolled in the collaborative were normatively aligned and invested in the overall research program and public good objectives. Therefore, it was inferred that only limited formal relationship building and maintenance activities were required, for these actors were all on the same page. Yet, there are myriad motivations for organisations to become involved in collaboratives, for example, delivering economic, organisational and/or societal benefits (see, e.g., Ankrah et al., 2013). A key part of building and maintaining stakeholder relationships relates to understanding and managing expectations, as well as a thorough understanding of the level of organisational 'buy-in'. Organisations that had been involved for many years were far more likely to have achieved sufficient internal agenda building to achieve on-ground changes, whereas organisations that only recently became part of the overall improved urban water management agenda were yet to build the requisite internal organisational support or commitment beyond accessing the latest science. Therefore, understanding the level of readiness to directly engage in delivering a change in practice is important for the broadening of experimentation, but this becomes significantly more challenging when working with more than eighty individual stakeholders ranging from universities, state government policy agencies and engineering consultancies to local municipalities.

6.4.3 Multiple Learning Opportunities

Research suggests that the mobilisation of key actor groups through continuous and persistent interaction is required to support knowledge transfer and embed new practices (Crona and Bodin, 2006; Spath and Rohracher, 2012). Within the context of interdisciplinary research, having numerous opportunities to engage with new research insights – particularly those that may not directly align with an individual's own disciplinary understandings and tacit experiences – assisted with broadening the reach and profile of the work. The experiences covered over two decades highlight a broadening of the mechanisms used for engaging with end users, from traditional means of producing guidelines and research reports, to incorporating collaborative design features that were explicitly interactive, engaging and supporting of open dialogues. Indeed, the design, variety and number of different knowledge exchange platforms was regarded as important for nurturing stakeholder relationships as well as in helping craft a broad narrative for change (Bos and Farrelly, 2015). The strategic embedding of multiple and varying interactive arenas appears to have assisted with going beyond experimentation to influence broader sectoral policy and practice change – evidenced by the suite of municipal governments that have altered their internal strategic and policy documentation as well as on-ground practices (see Bos and Farrelly, 2015).

This underscores the importance of generating a shared learning agenda among stakeholders to stimulate a broader, more reflexive learning culture (Bos et al., 2013) and providing diverse mechanisms to underpin this learning (i.e. workshops, training, conferences, etc.). Aligned with this is the need for dedicated evaluation activities within and across experiments and collaborative activities to help capture the outcomes (see Luederitz et al., 2016). However, too often evaluation remains an afterthought, or is superficial; thus to help move beyond experimentation, evaluation would ideally need to be undertaken both within the collaborative to generate insights regarding whether program design, delivery and processes were achieving their desired purpose (and allows space for adjusting the narrative should this be required) and capture causal information related to attribution regarding the outcomes achieved.

6.5 Persistent Challenges for Experimentation within Research-Industry Collaboration

Although this chapter has largely presented a positive change over the course of transitioning to more sustainable urban stormwater management, this hasn't been without significant and protracted challenges. Any form of engagement between and among a range of different stakeholders involves transactions costs. These are often amplified when working across traditional disciplinary boundaries and with non-scientific stakeholders and when industry and research time frames do not align. Being aware of and sympathetic to this context is important, for as identified in Bos et al. (2013), although the first few years of research interaction yielded what appeared to be limited scientific insights, once data emerged and began to 'tell a story', the perception of these transaction costs greatly diminished. Nonetheless, as the number of actors involved increases, managing the expectations of a wide array of stakeholders regarding timely objectives and outcomes remains a significant challenge.

Investing in relationship-building and stakeholder engagement requires time and effort (which increases as the density of the network increases), yet opportunities to do this are often limited due to unsympathetic institutional pressures. Nonetheless, having learnt from the experiences of researchers involved in early collective action platforms, contemporary researchers are beginning to embrace a more 'open and flexible' approach to sharing preliminary research insights and learning to respond to changing pressures (i.e. political, economic and regulative) (Brown et al., 2013). From an industry standpoint, it is well acknowledged that key organisational representatives are required to process, translate and communicate the research insights within their organisation, yet this remains a challenging interface when there are competing internal priorities. Indeed, Morison, Brown and Cocklin (2010) point out that due to a variety of persistent organisational constraints, the adoption of more sustainable urban stormwater management approaches has progressed but remains far from consistent across the metropolitan region.

Finally, as collaboratives shift away from the traditional linear innovation diffusion pathway towards embracing knowledge co-production through non-linear, recursive learning, interactive platforms, where actors can move in an out of the system, issues related to attribution become increasing more challenging and political. This could be addressed by embedding monitoring and evaluation within the design of the research collaborative, such that it tracks the process-related outcomes that are likely to deliver more sustained change and potential impact (Lang et al., 2012; Wiek et al., 2014; Luederitz et al., 2016).

6.6 Conclusion

There are many different ways of conceiving experimentation. Scholars are calling for research to go beyond examining technological experimentation towards advancing our understanding of governance experimentation (see, e.g., Kern and Howlett, 2009; Bos and Brown, 2012, Turnheim et al., Chapter 1). This chapter has embraced the opportunity to (re)examine technological experimentation by examining how, over a period of two decades, collaboratives have adapted, evolved and reshaped the type and extent of experimentation within the urban water sector of Australia. We outlined how the purpose of research-industry experimentation broadened from purely scientific/technological 'hypothesis-testing' experiments towards 'design-testing' experimentation and ultimately 'learning-by-doing' approaches. The multiple conceptualisations of experimentation were orchestrated by the bridging and intermediary work facili-tated by evolving collaborative platforms and key innovation champions within research and industry networks. What emerged were collaborative platforms that were actively supporting and enacting multiple conceptualisations of experimen-tation, via strong relational capacities, multiple interactive learning opportunities and strong actor networks. The interactions and engagement among researchers, industry and government actors, working within 'safe spaces', generated oppor-tunities for learning-by-doing and ultimately helped facilitate the broadening and embedding of new urban water management practices to address (and accommodate) climatic variability. These recursive collaborative forums helped actors craft new narratives to reflect new insights and understandings being generated from the research. Thus, shifting the narrative from stormwater as an end-of-pipe problem, to a potential resource, to a fundamental feature of a 'water sensitive city'.

Overall, simultaneous experimentation with different types of experiments (i.e. technical and social-interactive), within a broad actor network, is likely to support an expedited pathway to embedding new practices and governance approaches. Thus, these governance factors are important design considerations for future research-industry collaborations looking to generate broad sustainability outcomes by going beyond experimentation.

References

Ankrah, S. N., Burgess, T. F., Grimshaw, P., and Shaw, N. E. (2013). Asking both university and industry actors about their engagement in knowledge transfer: What single-group studies of motives omit. *Technovation*, 33(2), 50–65.

Barnett, J., and O'Neill, S. (2010). Maladaptation. *Global Environmental Change*, 20, 211–213.

Bettini, Y., Rijke, J., Farrelly, M. A., and Brown, R. R. (2013). Connecting levels and disciplines: Connective capacity of institutions and actors explored. In Edelenbos, J., Bressers, N., and Scholten, P. (eds.), *Water Governance as Connective Capacity*. The Netherlands: Ashgate, 129–149.

Bos, J. J., and Brown, R. R. (2012). Governance experimentation and factors of success in socio-technical transitions in the urban water sector. *Technological Forecasting and Social Change*, 79, 1340–1353.

Bos, J. J., Brown, R. R., and Farrelly, M. A. (2013). A design framework for creating social learning situations. *Global Environmental Change*, 23, 398–412.

Bos J. J., Brown R. R., and Farrelly, M. A. (2015). Building networks and coalitions to promote transformational change: Insights from an Australian urban water planning case study. *Environmental Innovation and Societal Transitions*, 15, 11–25.

Bos, J. J., and Farrelly, M. A. (2015). *Industry Impact of the CaWSC Research Program: Insights from Five Case Studies*. Melbourne, Australia: Cooperative Research Centre for Water Sensitive Cities.

Boardman, C., and Grey, D. (2010) The new science and engineering management: cooperative research centres as government policies, industry strategies, and organizations. *The Journal of Technology Transfer*, 35, 445–459.

Brown, R. R., Ashley, R., and Farrelly, M. A. (2011). Political and professional agency entrapment: An agenda for urban water research. *Water Resources Management*, 25, 4037–4050.

Brown, R. R., and Farrelly, M. A. (2009). Delivering sustainable urban water management: A review of the hurdles we face. *Water, Science and Technology*, 59, 839–846.

Brown, R. R., Farrelly, M. A., and Loorbach, D. A. (2013). Actors working the institutions in sustainability transitions: The case of Melbourne's stormwater management. *Global Environmental Change*, 23, 701–718.

Brown, R. R., Keath, N., and Wong, T. H. F. (2009). Urban water management in cities: Historical, current and future regimes. *Water, Science and Technology*, 59, 455–468.

Brown, R. R., Wong, T. H. F., and Deletic, A. (2015). Interdisciplinarity: How to catalyse collaboration. *Nature*, 525, 315–317.

Cooperative Research Centre for Water Sensitive Cities (CRCWSC) (2013). Annual Report 2012/2013. Melbourne, Victoria: Cooperative Research Centre for Water Sensitive Cities.

Crona, B. I., and Bodin, Ö. (2006). What you know is who you know? Communication patterns among resource extractors as a prerequisite for co-management. *Ecology and Society*, 11, 7.

Crona, B. L., and Parker, J. N. (2012). Learning in support of governance: Theories, methods and a framework to assess how bridging organizations contribute to adaptive resource governance. *Ecology and Society*, 17, 32.

Dentoni, D., and Bitzer, V. (2015). The role(s) of universities in dealing with global wicked problems through multi-stakeholder initiatives. *Journal of Cleaner Production*, 106, 68–78.

Etzkowitz, H., and Leydesdorff, L. (2000). The dynamics of innovation: From national systems and 'Mode 2' to a triple helix of university–industry–government relations. *Research Policy*, 29, 109–123.

Evans, J., and Karvonen, A. (2014) 'Give me a laboratory and I will lower your carbon footprint' – Urban laboratories and the governance of low-carbon futures. *International Journal of Urban and Regional Research*, 38, 413–430.

Farrelly, M., and Brown, R. R. (2011). Rethinking urban water management: Experimentation as a way forward? *Global Environmental Change*, 21, 721–732.

Ferguson, B., Brown, R. R., and Deletic, A. (2013). The enabling institutional context for integrated water management: Lessons from Melbourne. *Water Research*, 47, 7300–7314.

Fletcher, T. D., and Deletic, A. (2006). *A Review of Existing Water Quality Knowledge to Inform the Development of Melbourne Water's Waterways Water Quality Strategy*. Melbourne: Melbourne Water Corporation.

Folke, C., Hahn, T., Olsson, P., and Norberg, J. (2005). Adaptive governance of social–ecological systems. *Annual of Review Environment and Resources*, 30, 441–473.

Funtowicz, S., and Ravetz, J. (1993). Science for the post-normal age. *Futures*, 25, 735–755.

Gibbons, M., Limoges, C., Nowotny, H., Schwartzman, S., Scott, P., and Trow, M. (1994). *The New Production of Knowledge: The Dynamics of Science and Research in Contemporary Societies*. London: Sage.

Grin, J., Rotmans, J., and Schot, J. (2010) *Transitions to Sustainable Development: New Directions in the Study of Long Term Transformative Change*. London: Routledge.

Head, B. (2008). Three lenses of evidence-based policy. *Australian Journal of Public Administration*, 67, 1–11.

Huitema, D., and Meijerink, S. V. (2009). Water transitions, policy entrepreneurs and change strategies: Lessons learned. In Huitema, D., and Meijerink, S. V. (eds.), *Water Policy Entrepreneurs: A Research Companion to Water Transitions around the Globe*. Cheltenham, UK: Edward Elgar.

Kampelmann, S., Van Hollebeke, S., and Vandergert, P. (2016). Stuck in the middle with you: The role of bridging organisations in urban regeneration. *Ecological Economics*, 129, 82–93.

Kern, F., and Howlett, M. (2009). Implementing transition management as policy reforms: A case study of the Dutch energy sector. *Policy Sciences*, 42, 391–408.

Kivimaa, P. (2014) Government-affiliated intermediary organisations as actors in system-level transitions. *Research Policy*, 43, 1370–1380.

Lang, D. J., Wiek, A., Bergmann, M., Stauffacher, M., Martens, P., Moll, P., Swilling, M., and Thomas, C. J. (2012). Transdisciplinary research in sustainability science: Practice, principles, and challenges. *Sustainability Science*, 7, 25–43.

Lloyd, S., Wong, T., and Porter, B. (2002). The planning and construction of an urban stormwater management scheme. *Water Science and Technology*, 45, 1–10.

Loorbach, D. (2010). Transition management for sustainable development: A prescriptive, complexity-based governance framework. *Governance*, 23, 161–183.

Luederitz, C., Schapke, N., Wiek, A., Lang, D., Bergmann, M., Bos, J. J., Burch, S., et al. (2016). Learning through evaluation: A tentative evaluative scheme for sustainability transitions experiments. *Journal of Cleaner Production*. http://dx.doi.org/10.1016/j.jclepro.2016.09.005

Milly, P. C. D., Belancort, J., Falkenmark, M., Hirsch, R. M. Kendzewicz, Z. W. Lettenmaier, D. P., and Stouffler, R. (2008). Stationarity is dead: Whither water management? *Science*, 319, 573–574.

Morison, P. J., Brown, R. R., and Cocklin, C. (2010). Transitioning to a waterways city: Municipal context, capacity and commitment. *Water Science and Technology*, 62, 162–171.

Nevens, F., Frantzeskaki, N., Gorissen, L., and Loorbach, D. (2014). Urban transition labs: Co-creating transformative action for sustainable cities. *Journal of Cleaner Production*, 50, 111–122.

Olsson, P., Folke, C., and Hughes, T. P. (2008). Navigating the transition to ecosystem-based management of the Great Barrier Reef, Australia. *Proceedings of the National Academy of Sciences of the United States of America*, 105(28), 9489–9494. http://dx.doi.org/10.1073/pnas.0706905105

Perry, B. (2006). Science, society and the university: A paradox of values. *Social Epistemology*, 20, 201–19.

Ranga, M., and Etzkowitz, H. (2013). Triple helix systems: An analytical framework for innovation policy and practice in the Knowledge Society. *Industry and Higher Education*, 27, 237–262.

Rip, A. (1995). Introduction of new technology: Making use of recent insights from sociology and economics of technology. *Technological Analysis and Strategic Management*, 7, 417–31.

Spath, P., and Rohracher, H. (2012). Energy regions: The transformative power of regional discourses on socio-technical futures. *Research Policy*, 39, 449–458.

Taylor, A., Cocklin, C., Brown, R., and Wilson-Evered, E. (2011) An investigation of champion-driven leadership processes. *The Leadership Quarterly,* 22, 412–433.

Trencher, G. P., Yarime, M., and Kharrazi, A. (2013). Co-creating sustainability: Cross-sector university collaborations for driving sustainable urban transformations. *Journal of Cleaner Production*, 50, 40–55.

Trencher, G., Yarime, M., McCormick, K. B., Doll, C. N., and Kraines, S. B. (2014). Beyond the third mission: Exploring the emerging university function of co-creation for sustainability. *Science and Public Policy*, 41, 151–179.

Wiek, A., Talwar, S., O'Shea, M., and Robinson, J. (2014). Toward a methodological scheme for capturing societal effects of participatory sustainability research. *Research Evaluation*, 23, 117–132.

Williams, P. (February 2010). Special agents: The nature and role of boundary spanners. Presentation at the ESRC Seminar Series on Collaborative Futures: New Insights from Intra and Inter-Sectoral Collaborations, University of Birmingham.

Wong, T. H. F., and Brown, R. R. (2009). The water sensitive city: Principles for practice. *Water Science and Technology*, 60, 673–682.

Part II

Beyond Experiments

Transforming Climate Governance

7

Developing Transformative and Orchestrating Capacities for Climate Governance Experimentation in Rotterdam

KATHARINA HÖLSCHER, NIKI FRANTZESKAKI AND DERK LOORBACH

7.1 Introduction

As test beds for innovation, cities have become to be viewed as a vital part of the global search for solutions to sustainability and resilience in the face of problems such as climate change (Rosenzweig et al. 2011; IPCC, 2014; Chan et al., 2015). Cities embrace climate experimentation as a particular governance mode through which public and private actors search for and try out new solutions for tackling climate change (Castán Broto and Bulkeley, 2013; McCormick et al., 2013). Through learning-by-doing and contesting existing regulations and practices, it is suggested that experimentation may be more dynamic than traditional governance processes and better suited to dealing with uncertainty and contributing to the radical changes necessary to achieve resilience and sustainability in cities (Nevens, Frantzeskaki and Gorissen, 2013; Karvonen, Evans and van Heur, 2014; Evans, 2016). It can take place on different governance levels and it manifests in different types of innovations that test new policies, approaches and practices (Anguelovski and Carmin, 2011; Castán Broto and Bulkeley, 2013).

We conceptualise urban climate experimentation as a deliberate process of knowledge production for new approaches and solutions to facilitate climate-resilient sustainability transitions in cities. Currently, increasing attention is directed towards questions of what lies beyond experiments (Turnheim, Kivimaa and Berkhout, Chapter 1; Kivimaa et al., 2017). This concerns issues regarding what type of practices and processes result from experimentation and what impacts they bring about in urban governance processes and practical solutions. The interest resonates empirical observations that experimentation is often practiced unconnected to contemporary urban policy and planning processes (van Buuren et al., Chapter 8). As a result, innovations produced or nurtured through experimentation remain isolated stand-alone initiatives; their (positive and/or negative) contributions are neither sufficiently accounted for nor translated into mainstream practice (Ansell and Bartenberger, 2016; Evans 2016). This renders urban climate experimentation paradoxical: albeit employed as a contemporary mode of governance, it neither informs nor aligns with ongoing planning processes.

In this chapter, we examine how actors in the city of Rotterdam, the Netherlands, engage in climate experimentation to contribute to climate-resilient sustainability transitions. Climate-resilient sustainability transitions mark non-linear, radical changes in the structures, cultures and practices of a societal system to mitigate and adapt to climate change while promoting co-benefits for sustainability (Romero Lankao and Doman, 2011; Shaw et al., 2014; Chelleri et al., 2015). Against this background, whether and how cities succeed in 'moving beyond experimentation' is determined by the extent to which experimentation provides impulses to radical change that are embedded in and strategically connected with ongoing policy and planning processes. The aspiration to inform and acquire new ways of problem-solving implies some sort of learning from experimentation, especially about the viability, replicability and scalability of the experiment, as well as the translation of knowledge generated from the experiment to its broader context (Raven, van den Bosch and Weterings, 2010; Bos and Brown, 2012; Luederitz et al., 2016). Frameworks for assessing the impact of experiments appear as promising heuristics to map the impact and the directionality of innovations produced in the context of a strategy or goal of a city (Luederitz et al., 2016). With experiments operating in isolation or 'islands of innovation', cities that invest in experimentation need to examine and capitalise the 'value' these experiments bring in the policy mix in respect to the resources utilised (Frantzeskaki, Wittmayer and Loorbach, 2014; van Buuren et al., Chapter 8).

Actors in Rotterdam have a strong legacy of engaging in urban climate experimentation. They have formulated new strategic programmes that integrate climate change, sustainability and resilience objectives, set up new cross-departmental and public-private partnerships for the coordination of experimentation and implemented pilot projects that showcase innovative solutions. We apply a framework of transformative and orchestrating capacities to analyse different types of processes by which actors in Rotterdam engage in urban climate experimentation and to reflect on the extent to which experimentation contributes to a climate-resilient sustainability transition.

We conduct an in-depth case study (cf. Yin, 2003) that enables detailed observations to describe and explain the governance processes for experimentation in Rotterdam. The research findings are derived from a synthesis of data harvesting methods: (1) twenty-eight semi-structured interviews were conducted with climate governance practitioners in Rotterdam from different sectors (e.g. water and energy) and different societal groups (local, regional and national governmental authorities and public bodies, knowledge institutions, architectural offices, non-governmental organizations [NGOs] and community groups); (2) desk research was performed including a review of scientific articles, policy documents and media articles that pertain to the case; (3) two of the authors were involved in different vision and strategy development processes in Rotterdam; and (4) additional interviews and workshops were conducted in the course of the authors' involvement in climate governance processes in Rotterdam.

7.2 Governance Capacities for Urban Climate Experimentation

We apply a conceptual framework of governance capacities to examine how urban climate experimentation can contribute to climate-resilient sustainability transitions. Resonating an understanding of governance as encompassing the processes through which actors are able to collaborate to achieve common objectives (Stoker, 1998), the notion of governance capacity refers to the collective abilities of actors to engage in processes that work to achieve desired system-level outcomes, such as novel system structures or practices or a higher level of resilience.

For urban climate experimentation to contribute to climate-resilient sustainability transitions, governance actors require new types of capacities. We distinguish between transformative and orchestrating capacities. Transformative capacity processes aim at novelty creation as well as their anchoring and embedding in a specific context. Orchestrating capacity processes serve to coordinate multi-actor processes across scales and sectors to facilitate and align experimentation.

Our conceptualisation and operationalisation of the governance capacities draw on different scientific literatures. Climate governance scholars explore processes of urban climate experimentation and questions of embedding (Bulkeley and Castán Broto, 2013; Kivimaa et al., 2017) as well as of coordinating climate governance, resilience and sustainability across scales (Shaw et al., 2014; Chan et al., 2015). Sustainability transitions and resilience approaches emphasise experimentation as a means to challenge the status quo by intervening in complex and uncertain system dynamics that involve persistence, surprise and radical change (Frantzeskaki, Loorbach and Meadowcroft, 2012; Galaz et al., 2012; Loorbach, 2014; Olsson, Galaz and Boonstra, 2014). Meta-governance literature provides additional insights into instruments for coordination to facilitate alignment, collaboration and self-organisation in complex governance systems (Jessop, 1998; Sørensen, 2006; Kooiman and Jentoft, 2009; Capano, Howlett and Ramesh, 2015).

We reviewed the literatures regarding the processes they identify as manifest in transformative and orchestrating capacities. We inductively clustered these processes into the subfunctions each capacity performs to attain a desired system-level outcome (Table 7.1).

7.2.1 Transformative Capacity

Transformative climate governance includes processes to facilitate and embed innovation to contribute to systemic change. The growing attention to transformative capacity in diverse research fields builds on the recognition that radical changes are required to move towards sustainability and avoid possibly disastrous consequences on societal well-being and the environment (Wolfram, 2016). Sustainability transitions, resilience and climate governance literatures endorse the development of new ideas, practices, policies and approaches to test, embed, viable alternatives and eventually to transform to established institutions, infrastructures, behaviours, economies, etc. (Berkhout, Verbong and Wieczorek, 2010; Frantzeskaki, Loorbach and Meadowcroft, 2012; Castán Broto and Bulkeley, 2013; Westley et al., 2013).

We define transformative capacity as the collective abilities of actors to create novelties and to embed them in mainstream structures, cultures and practices to contribute to systemic change. Much emphasis has been paid to processes that enable *novelty creation*, which resonates the most conceptualised processes of experimentation. These include the provision of protected space for innovation (Raven et al., 2010; Bulkeley and Castán Broto, 2013), formation of informal networks (Olsson et al., 2006; Nevens et al., 2013) and leadership (Farrelly and Brown, 2011; Westley et al., 2013). To challenge dominant regimes, the innovation needs to gain traction and support (Raven et al., 2010; Westley et al., 2013). Processes to *anchor the novelty* include forming and supporting growing networks and alliances and developing common visions (Frantzeskaki et al., 2012; Nevens et al., 2013; Pahl-Wostl et al., 2013) that motivate and connect to other actors and ongoing processes to increase the traction and acceptance of the innovation and to encourage wider uptake (Elzen, Leeuwis and van Mierlo, 2008; Loorbach, 2010, 2014; Frantzeskaki et al., 2012; Chan et al., 2015; see also Carvalho and Lazzerini, Chapter 3). For an innovation to achieve wider uptake and result in more enduring change, processes for *novelty embedding* that make the implications and lessons from an innovation more generalisable are required to enable more widespread uptake and to fit them into the existing system (van Buuren et al., Chapter 8; Nevens et al., 2013). Embedding can take different forms, including the institutionalisation of new positions or regulations, replicating or scaling a specific project or mobilising experiments to reconfigure urban regimes (Elzen et al., 2008; Castán Broto and Bulkeley, Chapter 4; Bos and Brown, 2012).

Table 7.1. *Transformative and orchestrating capacities and processes*

Governance capacity and subfunctions		Actor processes and literature sources
Transformative capacity	*Novelty creation*	* Provision of protected spaces to nurture innovation in informal and heterogeneous (shadow) networks (Loorbach, 2010; Raven, van den Bosch and Weterings, 2010; Frantzeskaki, 2012; Smith and Raven, 2012; Bulkeley and Castán Broto, 2013; Sengers and Raven, 2015) * Leadership (by example) (Folke et al., 2005; Olsson et al., 2006; van Buuren and Loorbach, 2009; Farrelly and Brown, 2011; Bos and Brown, 2012; Marshall et al., 2012; Westley et al., 2013; Burch et al., 2014) * Informal and heterogeneous (shadow) networks that develop and test innovation and experiments (Folke et al., 2005; Olsson et al., 2006; Loorbach, 2010; Rotmans and Loorbach, 2010; Westley et al., 2013)
	Anchoring novelty	* Leadership and front runners who inspire and direct change and mobilise for support (Olsson et al., 2006; Gupta et al., 2010; Loorbach and Rotmans, 2010; Frantzeskaki et al., 2012; Westley et al., 2013;) * Heterogeneous networks with cross-scale and cross-sector connections create support for novelty (Loorbach, 2010; Wilson et al., 2013) * Alliances and advocacy networks forge to create broad acceptance of innovations (Olsson et al., 2006; Moore and Westley, 2011; Moore et al., 2011) * Shared future visions to provide guidance and inspiration (Loorbach and Rotmans, 2010; Nevens et al., 2013; Pahl-Wostl et al., 2013) * Perspectives of actors are reframed (Folke et al., 2005; Seyfang and Smith, 2007; Raven et al., 2010; Cumming et al., 2012)
	Novelty embedding	* Anticipating and recognising opportunities for change and crises (Olsson et al., 2006; Walker et al., 2009; Raven et al., 2010; Rotmans and Loorbach, 2010)

Table 7.1. (*cont.*)

Governance capacity and subfunctions		Actor processes and literature sources
		* Aligning institutional processes and structures to new vision and actions to increase implementation of novelty (Raven et al., 2010; Bos and Brown, 2012; Pahl-Wostl et al., 2013; Rijke et al., 2013; Burch et al., 2014; Bettini, Brown and de Haan, 2015; Wamsler, 2015)
		* Aligning short-term actions to long-term thinking and visions (Frantzeskaki et al., 2012; Nevens et al., 2013)
Orchestrating capacity	Strategic alignment	* Defining a shared and integrative strategic direction and reference points for governance (shared goals, vision, narrative) (Folke et al., 2005; Sørensen, 2006, 2014; Gupta, et al., 2010; Hodson and Marvin, 2010; Westley et al., 2013; Zonneveld and Spans, 2014; Abbott et al., 2015; Wittmayer et al., 2015)
		* Involving and enlist heterogeneous actor groups to create ownership over strategic direction (Frantzeskaki, 2014; Chan et al., 2015; Scourfield, 2015)
		* Linking to ongoing processes (Chan et al., 2015; Loorbach et al., 2015)
	Mediating across scales and sectors	* Recognising, brokering and integrating resources (financial, knowledge, human, etc.) and goals (Olsson et al., 2006; Hodson and Marvin, 2010; Vabo and Røiseland, 2012; Frantzeskaki et al., 2014; Friend et al., 2014; Beisheim and Simon, 2015; Wamsler, 2015)
		* Building trust, conflict management (Folke et al., 2005; Klijn and Edelenbos, 2007; Sørensen and Torfing, 2009; Jessop, 2011; Moore and Westley, 2011; Frantzeskaki et al., 2014; Wittmayer et al., 2014; Rauschmayer, Bauler and Schäpke, 2015; Scourfield, 2015)
		* Creating formal and informal spaces to exchange knowledge and manage conflicts (Loorbach, 2014)
		* Setting up connection nodes/adapting network structure to mediate knowledge and resources, optimise interactions and link formal and informal processes (Sørensen, 2006, 2014;

Table 7.1. (*cont.*)

Governance capacity and subfunctions	Actor processes and literature sources
	Chapin et al., 2010; Hodson and Marvin, 2010; Jessop, 2011; Moore et al., 2011; Frantzeskaki et al., 2014; Beisheim and Simon, 2015; Scourfield, 2015)
Creating opportunity contexts	* Providing financial incentives, regulations and institutional designs (Sørensen, 2006, 2014; Jessop, 2011; Vabo and Røiseland, 2012; Zonneveld and Spaans, 2014; Beisheim and Simon, 2015) * Determining (normative) action mandates and prioritising action and fields (Kooiman and Jentoft, 2009; Jessop, 2011; Chan et al., 2015) * Establishing and incorporating long-term and multi-scale thinking into decision-making and implementation processes (Chapin et al., 2010; Folke et al., 2010; Nevens et al., 2013; Frantzeskaki et al., 2014; Loorbach et al., 2015) * Understanding opportunity context and timing to create links between actors, strategies and resources and identify (new) resource opportunities (Westley et al., 2013; Frantzeskaki et al., 2014)

7.2.2 *Orchestrating Capacity*

The complex dynamics of persistent problems such as climate change require coordination of multi-scale, multi-sector and multi-actor processes, activities and goals (Galaz et al., 2012; Bulkeley and Betsill, 2013; Loorbach, 2014). Empirical observations of (the results of) urban experimentation processes reveal that these often remain isolated and unconnected to mainstream policy processes, manifesting for example in 'pilot paradoxes' (van Buuren et al., Chapter 8). To ensure concerted action for navigating climate-resilient sustainability transitions, meta-governance can serve to 'orchestrate' governance processes through facilitating and fostering processes of collaboration and coordination (Jessop, 1998; Sørensen, 2006, 2014; Capano et al., 2015). International organisations orchestrate climate actions across scales by aligning, enlisting and supporting state and subnational actors (Abbott et al., 2015; Chan et al., 2015). In cities, partnerships support the search for alternative solution spaces, bridge across sectors and scales, connect experiments to broader strategies and agendas and mediate knowledge and priorities (Hodson and Marvin, 2010; Cumming et al., 2012; Frantzeskaki et al., 2014).

We define orchestrating capacity as the collective abilities of actors to co-ordinate multi-actor processes across scales, sectors and time. Orchestrating processes serve the *strategic alignment* of actors and networks towards common goals that serve as reference points for concerted action through involving and enlisting heterogeneous actor groups and creating ownership (Sørensen, 2006, 2014; Hodson and Marvin, 2010; Scourfield, 2015). *Mediating across sectors and scales* includes processes that build trust and deal with conflicts (Klijn and Edelenbos, 2007; Beisheim and Simon, 2015). Connection nodes and communication channels support the exchange and dissemination of knowledge and resources (Scourfield, 2015). Finally, orchestrating includes processes to *create opportunity contexts* for concerted self-organisation through setting (political, financial and institutional) incentives and conditions that foster cooperation and win-win situations (Sørensen, 2006, 2014; Jessop, 2011).

7.3 Transformative and Orchestrating Capacities in Rotterdam

We first present a brief overview of the climate governance context in Rotterdam and then examine the processes through which actors in Rotterdam produce transforming and orchestrating capacities. We discuss what types of processes contribute to performing the capacities' subfunctions by linking our data to the theoretical framework. While we do not intend to stringently evaluate the performance in subfunctions, the focus on processes enables insights into what and how processes create opportunities and what are the challenges for performing the capacities' subfunctions. Table 7.3 summarises the findings from Rotterdam.

7.3.1 Climate Governance Architecture in Rotterdam

Rotterdam is an important delta city in the Netherlands, hosting the largest port in Europe and more than 650,000 inhabitants. Nearly 80 per cent of Rotterdam is located below the sea level (Molenaar, Dircke and Gebraad, 2013). Climate impacts include rising sea, river and groundwater levels, intense rainfalls and heat waves. Rotterdam faces additional socio-economic challenges related to social stratification, urban regeneration and the ongoing economic crisis that led to severe government budget cuts and withdrawals of public service provisions.

A variety of actors and networks engage in addressing climate change, sustainability and resilience in Rotterdam. The city government's cross-departmental Sustainability and Climate Adaptation offices spearhead these efforts in collaboration with other city departments such as the planning and economic policy departments. In addition, multiple regional and national policy and knowledge programmes (such as the Delta Programme, independent water boards and Knowledge for Climate) influence climate and sustainability governance in Rotterdam. The city government also participates in transnational city networks (e.g. C40, 100 Resilient Cities and Delta Cities) and public-private partnerships (e.g. the Rotterdam Climate Initiative and the Rotterdam Centre for Resilient Delta Cities) to promote knowledge production, sharing of good practices and collaboration.

Climate change was introduced on the city government's agenda in 2007 with ambitious CO_2-emission reduction targets and the launch of the Rotterdam Climate Initiative (Rotterdam Climate Initiative, 2007). Adjacent policy processes worked to reframe the view on water as a threat towards connecting water and adaptation to opportunities for building a more attractive city (de Greef, 2005). This was linked to the launch of the Rotterdam Climate Proof programme as part of the Rotterdam Climate Initiative in 2008 (Rotterdam Climate Initiative, 2009, 2012). The climate change focus was successively integrated with sustainability, liveability and resilience goals (Gemeente Rotterdam, 2012, 2015, 2016).

The 'Rotterdam approach' is marked by simultaneous innovations in strategic, institutional and operational approaches (Table 7.1). The latter focus especially on innovative and multifunctional ways to address climate adaptation while delivering co-benefits with other goals such as greening, recreation, community building and city planning (Figure 7.1). Another innovative project, the floating pavilion, showcases floating constructions (Figure 7.2).

Table 7.2. *Selected climate innovations that opened new approaches in Rotterdam*

Strategic innovations	Climate adaptation, sustainability and resilience strategies and programmes: new strategic goals and approaches that integrate different goals
Institutional innovations	Rotterdam Climate Initiative, Climate Adaptation and Sustainability Offices, etc.: new networks and partnerships to coordinate urban climate governance
Technical innovations	Benthemplein water square: combination of rainwater management with area development Zomerhofkwartier: climate-proof city quarter Boomjeskade: greening the waterfront Parking garage Museumplein: multifunctional water storage facility and parking garage Floating pavilion: floating construction to test building on water

Figure 7.1. The Benthemplein water square in use as a community square during a church service in May 2015.
Source: Reproduced with permission.

Figure 7.2. The Floating Pavilion, City Ports, Rotterdam.
Source: Photo courtesy of Niki Frantzeskaki, 2015.

7.3.2 *Transformative Capacity in Rotterdam*

Actors in Rotterdam were able to experiment with and develop new strategic approaches and implement pilot projects that tested novel practices and solutions. On a strategic level, actors from different city departments used political and international momentum to introduce mitigation goals and to reframe the city government's approach to water management from 'keeping water out' towards 'water as opportunity for liveability'. The latter was spurred by the International Architecture Biennale Rotterdam in 2005, which created space for actors from different city departments, regional public authorities and knowledge institutes to revisit in an informal process the fundamental challenges climate change will pose for Rotterdam. The involved actors emphasised that it was crucial to not work within the political context, but rather in safe surroundings. To attain wider support it was important to tell an enthusiastic and evidence-based story that combines a long-term perspective with small-scale practical solutions. The actors involved in this process were subsequently tasked to engage more formally in the development of the Rotterdam Water Plan 2 (Gemeente Rotterdam, 2007), taking up the new strategic approaches. The integrated goal formulation prompted new institutional structures in the city government (e.g. by setting up the cross-departmental Climate Office) and the establishment of new public-private partnerships.

On an operational level, the experimentation with innovative solutions has been an intrinsic part of the city's reinvention as a front runner in climate adaptation. The positioning of the projects as pioneering showcases helped bypass regulatory barriers (for example, through the recognition that building codes need to be adapted for building on water), attract support and raise financial resources. The development of the projects depended on highly committed and collaborating individuals from the city government, knowledge institutes and local businesses. For example, the Floating Pavilion Partnership was constituted by social, public and private actors with distinct roles for the planning, design, construction and research to create knowledge and resource synergies (Frantzeskaki et al., 2014). The construction and showcasing of the Floating Pavilion served to normalise the idea of building on water.

The experiences with strategic and operational experimentation spurred mostly informal learning. The actors involved in experimentation reflect on implementation processes and suitability, for example, regarding the need to involve local communities in project development. In this way experimentation in Rotterdam serves to achieve policy learning on how to co-create sustainable solutions for climate mitigation and adaptation. As a result, the way Rotterdam engages in experimentation starts to shift towards involving communities into taking measures and scaling innovative solutions from object-level to district-level. Regarding the latter, the architecture firm that implemented the Benthemplein water square sought to continue working with the knowledge they had gained and expand from the water square to a climate-proof city district. The process involved the local community to generate ideas and engage them in the implementation.

In Rotterdam, experimentation was successfully established as a viable planning and policy approach, and the experiences from experimentation resulted in a variety of learning outcomes. However, experimentation so far is practiced as a time- and energy-consuming add-on, and experimentation and its results are not taken up in ongoing governance processes. Rather than embedding the innovations in the city's planning approaches, tactical agendas and operating procedures, a central focus is on external communication to market Rotterdam as a front runner, build international networks and create 'export products'. There seems to be a trade-off between iconic projects and basic solutions that can be applied more widely. On the one hand, iconic projects such as the water square should serve to illustrate a principle of thinking about how to treat water management and climate adaptation in a more integral way. On the other hand, such understanding is not taken up in tactical programmes. There is a gap in the city government's middle management that bridges between strategic and operational levels to reflect on what has been done, what are the results and what are the implications for strategies and operational procedures.

7.3.3 Orchestrating Capacity in Rotterdam

In Rotterdam, orchestrating processes contribute to creating space for experimentation, connecting actors to each other for experimentation processes and mediating resources for experimentation. Actors from the city government's cross-departmental Sustainability and Climate Offices inhabit key positions in facilitating overarching strategy formulation and operational project implementation processes. The connection of different goals, including, for example, climate adaptation, liveability and economic development, provides a common framework for aligning (the different priorities of) actors for joint experimentation. For example, a project that seeks to develop the river as a tidal park connects seemingly opposing goals – economic activity, greening, biodiversity and recreation – and sparks collaboration between the port authority, the municipality and environmental organisations. Policy officers from the Sustainability and Climate offices are tasked with identifying opportunities for piggybacking on climate mitigation and adaptation initiatives by keeping an overview of how different projects come together, thinking of what can be combined and knowing whom to talk to. This helps identify and prioritise what actions are needed in which context and when.

Such processes of alignment and mediation also take place at different, more operational scales. For example, the Stadshavens Project Office facilitates and guides the redevelopment of the old city port area in Rotterdam. The office strategically connects different actors to joint visioning processes, develops and oversees vision-delivery implementation processes and monitors processes (Frantzeskaki et al., 2014). Actors involved in the planning of the climate-proof city quarter Zomerhofkwartier ensured connections to ongoing processes. This enabled

Table 7.3. *Opportunities and challenges for governance capacities in Rotterdam*

Governance capacity	Transformative capacity			Orchestrating capacity		
Subfunction	Novelty creation	Anchoring the novelty	Novelty embedding	Strategic alignment	Mediating across scales and sectors	Creating opportunity contexts
Opportunities	* Establishing protected spaces for strategic innovation and pilot projects * Framing pilot projects as showcasing to attract funding * Developing informal networks and coalitions of the willing * Mobilising personal commitment of innovation leaders	* Creating good visualisations and an inspiring story of vision and projects * Showing multiple benefits of innovation projects * Internationally advocating innovation projects * Creating economic spin-off opportunities	* Building on political support for innovation * Institutionalising strategies into new network structures * Showing benefits of new goals and projects to other (existing) priorities * Strategically selecting sites for innovative projects to facilitate up-scaling	* Defining an overarching strategic direction that creates co-benefits and links climate change to other goals * Integrating diverse knowledge sources and needs * Framing 'climate proofing' as key strategy for building on city assets * Involving diverse actors in strategy development to create ownership and collaboration	* Building trust and overcoming conflicts through knowledge partnerships * Setting up public-private partnerships for resource synergies * Establishing cross-departmental linkages and central nodes to concentrate climate change efforts * Brokering individual knowledge	* Integrating long-term time horizons in strategies and project plans * Experimenting with new funding schemes and pooling of resources * Creating economic spin-off opportunities from international networks and advocating * Identifying new funding sources (e.g. climate adaptation funds)
Challenges	* Mobilising budget and navigating regulatory	* Time constraints and small group of	* Diversifying localities from the city centre in view of equity	* Promoting an overarching city vision and integrating	* Limited involvement and outreach across	* Lacking expertise on approaches for long-term and

Table 7.3. (*cont.*)

Governance capacity	Transformative capacity			Orchestrating capacity		
Subfunction	Novelty creation	Anchoring the novelty	Novelty embedding	Strategic alignment	Mediating across scales and sectors	Creating opportunity contexts
	barriers for innovation * Integrating technical complexity in (increasing) [maintenance] costs	'ambassadors' limit outreach * Reaching out across departmental silos and sectors/organisations	* Translating between strategic and operational levels for institutional uptake of strategic goals and integrating lessons learned from pilot projects	broader goals (e.g. social equity) * Promoting ownership and take up of the strategic agenda across departments and actors in the city	governance levels, sectors and societal spheres * Time and resource constraints for communication and mediation * Limited awareness of and reaching out to civil society (organisations)	holistic business models and collaboration opportunities * Clarifying costs and benefits of measures and who pays and benefits

them to piggyback on the renewal of the sewage system: They could already tear out the tiles in that area to quickly express their idea for a rain garden.

Diverse knowledge creation processes support the alignment of actors, mediation of knowledge and identification of resource synergies for experimentation. These processes are facilitated by the Sustainability and Climate offices, public-private partnerships such as the Rotterdam Climate Initiative, the participation in transmunicipal city networks such as 100 Resilient Cities and other knowledge partnerships that act as intermediaries in disseminating knowledge and channelling information and ideas between different partners.

While the orchestrating processes at different (strategic, tactical and operational) levels help create space and identify opportunities for experimentation, they do not extend beyond a relatively small network of key actors from the city government, knowledge institutes and businesses. As a result, experimentation and its results remain unconnected from ongoing planning processes and there are no strategic endeavours to position experimentation in relation to other projects. For example, while the new strategic sustainability and resilience goals suggest a reorientation of the city's development, many other goals and activities still build on business-as-usual activity. Additionally, there are no formal context conditions for enabling experimentation and accounting for the co-benefits between different objectives addressed through experimentation to take up the results into mainstream action. Public and private actors engaged in experimentation highlight the need for more creativity regarding where to source finance and for clear guidelines regarding competing for funding. Throughout experimentation processes these actors achieve innovative financing often in an ad hoc way. A project that is currently starting between a local architecture company and housing corporation seeks to find out if the greening of roofs can be connected to the maintenance system of the houses to integrate it in a cost-effective way when something needs to be replaced.

7.4 Discussion

The concomitant experimentation processes at strategic and operational governance levels led to the ongoing innovation of strategies, institutional structures and solutions to address climate change, resilience and sustainability in Rotterdam. These activities were led by relatively small groups of actors from the city government, knowledge institutes and local businesses that organise mostly informally. The integration of diverse goals and priorities spurred new institutional structures, for example, cross-departmental offices and public-private partnerships, which in turn helped create space for experimentation and to connect actors and resources across scales and sectors.

The analysis of transformative and orchestrating capacities helps us discern two interrelated challenges for (moving beyond) experimentation in Rotterdam and for increasing the contribution of experimentation to a climate-resilient sustainability transition. Firstly, while there are a variety of experimentation processes to develop innovative strategies and projects, these largely serve to showcase innovative solutions and externally market Rotterdam as a front runner city in climate adaptation and sustainability. In Rotterdam, experimentation is used as a process to generate iconic innovations to mobilise resources and especially to draw (international) attention. This way of experimentation seems to underpin competitive urbanism without being connected to internal embedding processes. It also reveals the limitations of experimentation in light of contributing to systemic change; it is conforming to existing neo-liberal rationalities, hence protecting and extending market relations rather than fundamentally questioning them (Evans, 2016; Jhagroe, 2016). As neither experimentation nor sustainability transitions are value-free, the use of experimentation to create highly visible international showcases raises questions about the choices of where experiments are conducted, for what reasons (e.g. climate adaptation, mitigation and/or sustainability) and for whom (Evans, 2016).

Secondly, orchestrating capacity is produced by relatively small and informal groups of actors that strategically select sites for experimentation, mediate knowledge and connect actors and resources. This way of orchestrating (so far) resembles ad hoc processes that are hardly legitimised or structurally supported. While there seems to be a lot of strength in this hybrid approach, it is also limited with regard to its impact on wider policy and planning processes. Aligning experimentation processes and its results to a wider range of actors could help connect experiments to ongoing processes and projects and build a critical mass of innovation for systemic change (Hodson, Marvin and Bulkeley, 2013; Frantzeskaki et al., 2014). This might help overcome the 'pilot paradox', i.e. the disconnect between experimentation and formal planning and policy processes (van Buuren et al., Chapter 8). It could also close gaps between strategic and operational levels to translate strategic goals and lessons from pilot projects into tactical agendas and programmes, for example, by creating framework conditions for projects with long-term financial returns, strategically connecting experimentation processes to other projects and helping determine context-specific (combinations of) solutions from a diversity of measures. The climate change adaptation projects and the teams implementing them could also connect to other planning processes and programmes in Rotterdam that experiment in the social domain to create further synergies between goals.

Ironically, starting to engage in experimentation as a governance mode in Rotterdam can by itself be considered an instance of experimentation (Farrelly and Bos, Chapter 6). Transformative and orchestrating capacities came about from experimenting with innovative strategies that integrate a variety of priorities and goals and new institutional structures to coordinate multi-actor (experimentation) processes. They are consistently built through the informal learning of the involved actors in engaging in transformative and orchestrating processes.

7.5 Conclusion

In this chapter, we have analysed the transformative and orchestrating capacities that are manifest in the urban climate experimentation processes in the city of Rotterdam. We have argued that experimentation which contributes to climate-resilient sustainability transitions requires such types of capacities to make the generated and tested knowledge on new approaches and solutions productive for systemic change by connecting and aligning experimentation with ongoing planning processes in the city and embedding experiments in policy and planning.

Our analysis of transformative and orchestrating capacities for urban climate experimentation in Rotterdam reveals several opportunities and challenges for the role of experimentation in climate-resilient sustainability transitions. While actors in Rotterdam have successfully established experimentation as a new governance mode to address climate change, sustainability and resilience, this approach is mainly legitimised by marketing the city internationally as a front runner in climate adaptation. As a result, individual innovations remain mostly stand-alone initiatives that are not taken up or connected with other activities in the city. A key question is how to further embed and support transformative and orchestrating capacities to enhance the abilities of actors to position experimentation within other processes, mediate across (strategic and operational) governance levels and create (more formal) space for reflection on the lessons of experiments for financing structures and operating procedures.

Acknowledgements

This research was funded by the EU project IMPRESSIONS (Impacts and Risks from High-End Scenarios: Strategies for Innovative Solutions, www.impressions-project.eu). The authors would like to especially thank Bruno Turnheim and Paula Kivimaa and all participants of the Innovations in Climate Governance (INOGOV) workshop Beyond Experiments for their feedback on earlier versions of this chapter.

References

Abbott, K. W., Genschel, P., Snidal, D., and Zangl, B. (eds.) (2015). *International Organizations as Orchestrators*. Cambridge: Cambridge University Press.

Anguelovski, I., and Carmin, J. (2011). Something borrowed, everything new: Innovation and institutionalization in urban climate governance. *Current Opinion in Environmental Sustainability*, 3, 169–175.

Ansell, C. K., and Bartenberger, M. (2016). Varieties of Experimentalism. *Ecological Economics*, 130, 64–73. https://dx.doi.org/10.2139/ssrn.2475844

Beisheim, M., and Simon, N. (August 2015). Meta-governance of partnerships for sustainable development: Actors' perspectives on how the UN could improve partnerships' governance services in areas of limited statehood. Berlin: SFB-Governance Working Paper Series, No. 68.

Berkhout, F., Verbong, G., Wieczorek, A., Raven, R., Lebel, C., and Bai, X. (2010). Sustainability experiments in Asia: Innovations shaping alternative development pathways? *Environmental Science and Policy*, 13, 261–271.

Bettini, Y., Brown, R., and de Haan, F. J. (2015). Exploring institutional adaptive capacity in practice: Examining water governance adaptation in Australia. *Ecology and Society*, 20(1), 47. http://dx.doi.org/10.5751/ES-07291–200147.

Bos, J. J., and Brown, R. R. (2012). Governance experimentation and factors of success in socio-technical transitions in the urban water sector. *Technological Forecasting and Social Change*, 79(7), 1340–1353.

Bulkeley, H., and Castán Broto, V. (2013). Government by experiment? Global cities and the governing of climate change. *Transactions of the Institute of British Geographers*, 38(3), 361–375.

Bulkeley, H. A., and Betsill, M. M. (2013). Revisiting the urban politics of climate change. *Environmental Politics*, 22, 136–154.

Burch, S., Shaw, A., Dale, A., and Robinson, J. (2014). Triggering transformative change: A development path approach to climate change response in communities. *Climate Policy*, 14(4), 467–487. doi: 10.1080/14693062.2014.876342.

Capano, G., Howlett, M., and Ramesh, M. (2015). Bringing governments back in: Governance and governing in comparative policy analysis. *Journal of Comparative Policy Analysis: Research and Practice*, 17(4), 311–321. doi: 10.1080/13876988.2015.1031977.

Castán Broto, V., and Bulkeley, H. (2013). A survey of urban climate change experiments in 100 cities. *Global Environmental Change*, 23, 92–102.

Chan, S., Falkner, R., van Asselt, H., and Goldberg, M. (2015). Strengthening non-state climate action: A progress assessment of commitments launched at the 2014 UN Climate Summit. Centre for Climate Change Economics Policy, Working Paper No. 242. Grantham Research Institute on Climate Change and the Environment, Working Paper No. 216.

Chapin III, S. F., Carpenter, S. R., Kofinas, G. P., Folke, C., Abel, N., Clark, W. C., Olsson, P., Stafford Smith, D. M., Walker, B., Young, O. R., Berkes, F., Biggs, R., Grove, J. M., Naylor, R. L., Pinkerton, E., Steffen, W., and Swanson, F. J. (2010). Ecosystem stewardship: Sustainability strategies for a rapidly changing planet. *Trends in Ecology and Evolution*, 25(4), 241–249.

Chelleri, L., Waters, J. J., Olazabal, M., and Minucci, G. (2015). Resilience trade-offs: Addressing multiple scales and temporal aspects of urban resilience. *Environment & Urbanization*, 27(1), 181–198.

Cumming, G. S., Olsson, P., Chapin III, F. S., and Holling, C. S. (2012). Resilience, experimentation, and scale mismatches in social-ecological landscapes. *Landscape Ecology*. doi:10.1007/s10980-012–9725-4.

De Greef, P. (2005) (ed.). *Rotterdam Waterstad 2035*. Heijningen: Jap Sam Books.

Elzen, B., Leeuwis, C., and van Mierlo, B. (2008). Anchorage of Innovations: Assessing Dutch efforts to use the greenhouse effect as an energy source. Working Paper. Wageningen: Wageningen University.

Evans, J. (2016). Trials and tribulations: Problematizing the city through/as urban experimentation. *Geography Compass*, 10(10), 429–443.

Farrelly, M., and Brown, R. (2011). Rethinking urban water management: Experimentation as a way forward? *Global Environmental Change*, 21(2), 721–732.

Folke, C., Carpenter, S. R., Walker, B., Scheffer, M., Chapin, T., and Rockström, J. (2010). Resilience thinking: Integrating resilience, adaptability and transformability. *Ecology and Society*, 15(4), 20.

Folke, C., Hahn, T., Olsson, P., and Norberg, J. (2005). Adaptive governance of social-ecological systems. *Annual Review of Environment and Resources*, 30(1), 441–473.

Frantzeskaki, N., Loorbach, D., and Meadowcroft, J. (2012). Governing societal transitions to sustainability. *International Journal of Sustainable Development*, 15(1), 19–36.

Frantzeskaki, N., Wittmayer, J., and Loorbach, D. (2014). The role of partnerships in 'realizing' urban sustainability in Rotterdam's City Ports Area, the Netherlands. *Journal of Cleaner Production*, 65, 406–417. doi: 10.1016/j.jclepro.2013.09.023.

Friend, R., Jarvie, J., Orleans Reed, S., Sutarto, R., Thinphanga, P., Canh Toan, V. (2014). Mainstreaming urban climate resilience into policy and planning; reflections from Asia. *Urban Climate*, 7, 6–19. doi: http://dx.doi.org/10.1016/j.uclim.2013.08.001.

Galaz, V., Crona, B., Österblom, H., Olsson, P., and Folke, C. (2012). Polycentric systems and interacting planetary boundaries: Emerging governance of climate change – ocean acidification – marine biodiversity. *Ecological Economics*, 81, 21–32. doi: 10.1016/j.ecolecon.2011.11.012.

Gemeente Rotterdam (2007). *Waterplan 2: Werken an water voor een aantrekkelijke en klimaatbestendige stad*. Rotterdam: Gemeente Rotterdam.

Gemeente Rotterdam (2012). *Programma Duurzaam, Investeren in duuzaame groei*. Rotterdam: Gemeente Rotterdam.

Gemeente Rotterdam (2015). *Duurzaam dichter bij de Rotterdammer; Programma Duurzaam 2015–2018*. Rotterdam: Gemeente Rotterdam.

Gemeente Rotterdam (2016). *Rotterdam resilience strategy: Ready for the 21st century*. Rotterdam: Gemeente Rotterdam. http://lghttp.60358.nexcesscdn.net/8046264/images/page/-/100rc/pdfs/strategy-resilient-rotterdam.pdf. Accessed 20 September 2016.

Gupta, J., Termeer, C. J. A. M., Klostermann, J. E. M., Meijerink, S., van den Brink, M. A., Jong, P., Nooteboom, S. G., Bergsma, E. J. (2010). The adaptive capacity wheel: A method to assess the inherent characteristics of institutions to enable the adaptive capacity of society. *Environmental Science & Policy*, 13:459–471.

Hodson, M., and Marvin, S. (2010). Can cities shape socio-technical transitions and how would we know if they were? *Research Policy*, 39, 477–485.

Hodson, M., Marvin, S., and Bulkeley, H. (2013). The intermediary organisation of low carbon cities: A comparative analysis of transitions in Greater London and Greater Manchester. *Urban Studies*, 50, 1403–1422.

International Panel on Climate Change (IPCC) (2014). Climate Change 2014: Impacts, Adaptation and Vulnerability. *IPCC Working Group II Contribution to AR5*. Summary for Policymakers. Cambridge: Cambridge University Press.

Jessop, B. (1998). The rise of governance and the risks of failure: The case of economic development. *International Social Science Journal*, 50(155), 29–45.

Jessop, B. (2011). Metagovernance. In Bevir, M. (ed.), *The Sage Handbook of Governance*. London: Sage, 106–123.

Jhagroe, S. (2016). Urban transition politics: How struggles for sustainability are (re) making urban spaces. PhD thesis, Erasmus University Rotterdam.

Karvonen, A., Evans, J. P., and van Heur, B. (2014). The politics of urban experiments: Radical change or business as usual? In Marvin, S., and Hodson, M. (eds.), *After Sustainable Cities*. London: Routledge, 105–114.

Kivimaa, P., Hildén, M., Huitema, D., Jordan, A., and Newig, J. (2017). Experiments in climate governance: A systematic review of research on energy and built environment transitions. *Journal of Cleaner Production*, 169: 17–29.

Klijn, E., and Edelenbos, J. (2007). Metagovernance as network management. In Sørensen, E., and Torfing, J. (eds.), *Theories of Democratic Network Governance*. New York: Palgrave Macmillan, 199–214.

Kooiman, J., and Jentoft, S. (2009). Meta-Governance: Values, norms and principles, and the making of hard choices. *Public Administration*, 87(4), 818–836.

Loorbach, D. (2010). Transition management for sustainable development: A prescriptive, complexity-based governance framework. *Governance: An International Journal of Policy Administration and Institutions*, 23(1), 161–183.

Loorbach, D. (2014). *To Transition! Governance Panarchy in the New Transformation*. Inaugural Lecture. Erasmus University Rotterdam.

Loorbach, D., and Rotmans, J. (2010). The practice of transition management: Examples and lessons from four distinct cases. *Futures*, 42(3), 237–246.

Loorbach, D., Frantzeskaki, N., and Huffenreuter, L. R. (2015). Transition management: Taking stock from governance experimentation. *Journal of Corporate Citizenship*, 58, 48–66.

Luederitz, C., Schäpke, N., Wiek, A., Lang, D. J., Bergmann, M., Bos, J. J., Burch, S., Davies, A., Evans, J., König, A., Farrelly, M. A., Forrest, N., Frantzeskaki, N., Gibson, R. B., Kay, B., Loorbach, D., McCormick, K., Parodi, O., Rauschmayer, F., Schneidewind, U., Stauffacher, M., Stelzer, F., Trencher, G., Venjakob, J., Vergragt, P. J., von Wehrden, H., and Westley, F. R. (2016). Learning through evaluation. A tentative evaluation scheme for sustainability transition experiments. *Journal of Cleaner Production* (2016). http://dx.doi.org/10.1016/j.jclepro.2016.09.005.

Marshall, N. A., Park, S. E., Adger, W. N., Brown, K., and Howden, S. M. (2012). Transformational capacity and the influence of place and identity. *Environmental Research Letters*, 7, 1–9.

McCormick, K., Anderberg, S., Coenen, L., and Neij, L. (2013). Advancing sustainable urban transformation. *Journal of Cleaner Production*, 50, 1–11.

Molenaar, A., Dircke, P., and Gebraad, C. (2013). Rotterdam. In Molenaar, A., Aerts, J., Dircke, P., and Ikert, M. (eds.), *Connecting Delta Cities: Resilient Cities and Climate Adaptation Strategies*. Rotterdam: City of Rotterdam, 30–51.

Moore, M.-L., and Westley, F. (2011). Surmountable chasms: Networks and Social innovation for resilient systems. *Ecology and Society*, 16(1), 5. www.ecology andsociety.org/vol16/iss1/art5/

Moore, M.-L., Westley, F. R., Tjornbo, O., and Holroyd, C. (2011). The loop, the lens and the lesson: Using resilience theory to examine public policy and social innovation. In Nicholls, A., and Murdock, A. (eds.), *Social Innovation: Blurring Boundaries to Reconfigure Markets*. Basingstoke: Palgrave Macmillan, 320.

Nevens, F., Frantzeskaki, N., Gorissen, L., and Loorbach, D. (2013). Urban transition labs: Co-creating transformative action for sustainable cities. *Journal of Cleaner Production*, 50, 111–122.

Olsson, P., Galaz, V., and Boonstra, W. J. (2014). Sustainability transformations: A resilience perspective. *Ecology and Society* 19(4), 1.

Olsson, P., Gunderson, L. H., Carpenter, S. R., Ryan, P., Lebel, L., Folke, C., and Holling, C. S. (2006). Shooting the rapids: Navigating transitions to adaptive governance of social-ecological systems. *Ecology and Society*, 11(1), 18. www.ecologyandsociety .org/vol11/iss1/art18

Pahl-Wostl, C., Becker, G., Knieper, C., and Sendzimir, J. (2013). How multilevel societal learning processes facilitate transformative change: A comparative case study analysis on flood management. *Ecology and Society*, 18(4), 58. http://dx.doi.org/10.5751/ES-05779-180458.

Rauschmayer, F., Bauler, T., and Schäpke, N. (2015). Towards a thick understanding of sustainability transitions: Linking transition management, capabilities and social practices. *Ecological Economics*, 109, 211–221.

Raven, R., van den Bosch, S., and Weterings, R. (2010). Transitions and strategic niche management: Towards a competence kit for practitioners. *International Journal of Technology Management*, 51(1), 57–74.

Rijke, J., Farrelly, M., Brown, R., and Zevenbergen, C. (2013). Configuring transformative governance to enhance resilient urban water systems. *Environmental Science & Policy*, 25(2013), 62–72.

Romero Lankao, P., and Dodman, D. (2011). Cities in transition: Transforming urban centers from hotbeds of GHG emissions and vulnerability to seedbeds of sustainability and resilience. *Current Opinion in Environmental Sustainability*, 3(3), 113–120.

Rosenzweig, C., Solecki, W. D., Hammer, S. A., and Mehrotra, S. (eds.) (2011). *Climate Change and Cities: First Assessment Report of the Urban Climate Change Research Network*. Cambridge: Cambridge University Press.

Rotmans, J., and Loorbach, D. (2010). Towards a better understanding of transitions and their governance: A systemic and reflexive approach. In Grin, J., Rotmans, J., and Schot, J. (eds.), *Transitions to Sustainable Development: New Directions in the Study of Long Term Transformative Change*. New York: Routledge, 105–220.

Rotterdam Climate Initiative (2007). Action programme and objectives 2007–2010. Rotterdam: City of Rotterdam.

Rotterdam Climate Initiative (2009). Rotterdam climate proof: The Rotterdam challenge on water and climate adaptation. 2009 adaptation programme. Rotterdam. Rotterdam: City of Rotterdam. www.google.nl/url?sa=t&rct=j&q=&esrc=s&source=web&cd= 4&cad=rja&uact=8&ved=0ahUKEwjT0Z71xYXRAhXROlAKHY6eAwIQFgg4M AM&url=http%3A%2F%2Fwww.rotterdamclimateinitiative.nl%2Fdocuments%2F 2015-en-ouder%2FRCP%2FEnglish%2FRCP_adaptatie_eng.pdf&usg=AFQjCNF WDPhuaDUNGD_W_o3LCMxp8laeOg. Accessed 10 June 2016.

Rotterdam Climate Initiative (2012). Rotterdam climate change adaptation strategy. Rotterdam: City of Rotterdam. www.rotterdamclimateinitiative.nl/documents/2015 -en-ouder/Documenten/20121210_RAS_EN_lr_versie_4.pdf. Accessed 10 June 2016.

Scourfield, P. (2015). Implementing co-production in adult social care: An example of meta-governance failure? *Social Policy and Society*, 14(4), 541–554. doi: 10.1017/ S1474746414000438.

Sengers, F., and Raven, R. (2015). Towards a spatial perspective on niche development: The case of Bus Rapid Transit. *Environmental Innovation and Societal Transitions*, 17, 166–182.

Seyfang, G., and Smith, A. (2007). Grassroots innovations for sustainable development: Towards a new research and policy agenda. *Environmental Politics*, 16(4), 584–603.

Shaw, A., Burch, S., Kristensen, F., Robinson, J., and Dale, A. (2014). Accelerating the sustainability transition: Exploring synergies between adaptation and mitigation in British Columbian communities. *Global Environmental Change*, 25, 41–51.

Smith, A., and Raven, R. (2012). What is protective space? Reconsidering niches in transitions to sustainability. *Research Policy*, 7, 41(6). doi: 10.1016/j.respol.2011.12.012.

Sørensen, E. (2006). Metagovernance: The changing role of politicians in processes of democratic governance. *Public Administration*, 36(1), 98–114. doi: 10.1177/0275074005282584.

Sørensen, E. (16–17 September 2014). The metagovernance of public innovation in governance networks. Paper presented at Policy and Politics conference, Bristol.

Sørensen, E., and Torfing, J. (2009). Making governance networks effective and democratic through metagovernance. *Public Administration*, 87(2), 234–258.

Stoker, G. (1998). Governance as theory: Five propositions. *International Social Science Journal*, 50(155), 17–28.

Vabo, S. I., and Røiseland, A. (2012). Conceptualizing the tools of government in urban network governance. *International Journal of Public Administration*, 35(14), 934–946. doi: 10.1080/01900692.2012.691243.

Van Buuren, A., and Loorbach, D. (2009). Policy innovation in isolation? Conditions for policy renewal by transition arenas and pilot projects. *Public Management Review*, 11(3), 375–392.

Walker, B. H., Abel, N., Anderies, J. M., and Ryan, P. (2009). Resilience, adaptability, and transformability in the Goulburn-Broken Catchment, Australia. *Ecology and Society*, 14(1), 12.

Wamsler, C. (2015). Mainstreaming ecosystem-based adaptation: Transformation toward sustainability in urban governance and planning. *Ecology and Society*, 20(2), 30.

Westley, F. R., Tjornbo, O., Schultz, L., Olsson, P., Folke, C., Crona, B., and Bodin, Ö. (2013). A theory of transformative agency in linked social-ecological systems. *Ecology and Society*, 18(3), 27. doi:10.5751/ES-05072–180327.

Wilson, S., Pearson, L. J., Kashima, Y., Lusher, D., and Pearson, C. (2013). Separating adaptive maintenance (resilience) and transformative capacity of social-ecological systems. *Ecology and Society*, 18(1), 22. http://dx.doi.org/10.5751/ES-05100–180122.

Wittmayer, J., Van Steenbergen, F., Loorbach, D., Mock, M,. Omann, I., and Kirner, B. (2014). Exploring the transformative potential of communities. In Wittmayer, J., Roorda, C., and Van Steenbergen, F. (eds.), *Governing Urban Sustainability Transitions – Inspiring examples*. Rotterdam: DRIFT, Creative Commons, 83–89.

Wolfram, M. (2016). Conceptualizing urban transformative capacity: A framework for research and policy. *Cities*, 51, 121–130. doi: 10.1016/j.cities.2015.11.011.

Yin, R. K. (2003). *Case Study Research: Design and Methods*. Thousand Oaks, CA: Sage Publications.

Zonneveld, W., and Spaans, M. (2014). Meta-governance and developing integrated territorial strategies: The case study of MIRT territorial agendas in the Randstad (Netherlands). *Planning Theory & Practice*, 15(4), 543–562.

8

The Pilot Paradox

Exploring Tensions between Internal and External Success Factors in Dutch Climate Adaptation Projects

ARWIN VAN BUUREN, HELEEN VREUGDENHIL, JITSKE VAN POPERING-VERKERK, GERALD JAN ELLEN, CORNIEL VAN LEEUWEN AND BAS BREMAN

8.1 Introduction

Climate change poses new, intricate challenges for public authorities. They have to develop and adjust policies to adapt to these changing conditions. To do so, novel ideas and innovative approaches are developed and tested in pilot projects (Kivimaa et al., 2017). Such projects represent a significant new source of innovation and capability formation (Berkhout et al., 2010).

Pilot projects are perceived and often acknowledged as a relatively easy and safe manner in which to explore new pathways to deal with consequences of climate change. Pilot projects help ascertain whether new solutions – for instance, for fresh water supply, flood protection or water retention – are effective before they are embedded in formal policies, implementation programmes (Bos, Brown and Farrelly, 2013), or broader policy mixes (Nair and Howlett, Chapter 9). Pilot projects function as a safe area to develop novel ways of thinking and allow for exploring creative methods and mobilising new sources of knowledge (Smith and Raven, 2012).

In this chapter, we consider pilot projects as innovation (technological, process and otherwise) projects conducted in the social-physical environment and oriented towards policy learning. They can be distinguished from experiments conducted in a laboratory, randomised control trials or even social experiments that focus on increasing system understanding (Vreugdenhil, 2010). Essentially, pilot projects are used to apply innovations in a real-world context. To take a next step in the course of an innovation process, a pilot project serves as a space where the understanding and the embedding of the invention in its physical and social context can be improved – an improvement that can typically not be obtained otherwise (Banks, 2009). Pilots function as 'fringe spaces for current dominant technologies or alternatives to currently used methods of governance' (Kivimaa et al., 2017: 2). Moreover, pilot projects can be used to support the implementation of new policies and to demonstrate, or convince others about, the use of the innovation (Vreugdenhil et al., 2010).

Pilot projects are usually organised at the boundaries of formal (public) organisations, or even outside the dominant policy regime (Rotmans and Loorbach, 2009). We define policy regime as the dominant policy paradigm and governance approach, the normal way of doing and the rules of the game within and between the various mother organisations of a pilot project (May and Jochim, 2013). Distance to these mother organisations and thus to the dominant policy regime is necessary to give the pilot some room for innovation and exploration (van Buuren and Loorbach, 2009). Furthermore, pilot projects can be distinguished from regular projects because of the availability of additional resources (staffing, expertise and funding) and application at a relatively small scale (e.g. at floodplain level instead of river stretch or at street level instead of a whole city).

The same characteristics that facilitate the development and implementation of a pilot project may cause a tension in the wider uptake of the pilot by the various mother organisations and thus the existing policy regime (Vreugdenhil, 2010). Nair and Howlett (Chapter 9) have found that political support and the need for synergies with ongoing policies and programmes are nearly always necessary for scaling up. In combination with effective pilot planning and strong monitoring and evaluation, these factors create a sufficient condition for the pilot uptake. Proximity dimensions and contextual developments are also of importance. Coenen, Raven and Verbong (2010) argue that (the place/location) where innovations occur is not coincidental and that pilots should not be considered in isolation but within larger temporal and spatial contexts. Similarly, Berkhout et al. (2010) discuss how sustainability experiments link to global knowledge and technology flows and how they have the ability to reshape emergent socio-technical regimes and so contribute to alternative development pathways. This implies that the uptake of pilot projects relates to the extent to which they 'fit' with their context. A 'misfit' with the policy context, limited political support, poor planning, and monitoring and evaluation may hinder the uptake. This contradicts the observations that pilots usually take place at the boundaries of policies and policy organisations (Rotmans and Loorbach, 2009), that pilot projects are not necessarily initiated for policy learning and that the role of evaluation is not well defined, leading to poor evaluation programmes (Sanderson, 2002). Hence, the ultimate result of a pilot in terms of generating policy change is often quite low (Vreugdenhil, 2010).

We therefore consider the existence of a 'pilot paradox': the conditions that are necessary to give a pilot room to experiment and to learn (and, thus, arrive at innovative results) also seem to constitute the main barriers to the broader uptake and translation of its results into changes in policy goals, content or instruments. Pilot results are often misaligned with prevailing policy paradigms and, because of the distance between the pilot and the mother organisations, the latter are ill-prepared to receive pilot results.

In this chapter, we explore this potential pilot paradox. We bring together six climate adaptation pilots in the Netherlands, evaluated between 2010 and 2016 (Vreugdenhil, 2010; Ellen and Ottow, 2012; Breman et al., 2014; van Buuren et al., 2015b). In three cases, we were involved as a researcher or expert and in the other three cases we were asked by governmental agencies to conduct evaluation studies.

The chapter is structured as follows. In the first two sections, we explore the conditions for successful experimentation in pilots and the necessary conditions for policy change and learning in existing policy regimes. To clarify what we mean by success in this chapter, we distinguish between internal and external success. Conditions for internal success are discussed in Section 8.2 and conditions for external success in Section 8.3. In Section 8.4, we confront these two sets of conditions to illustrate the pilot paradox. In Section 8.5, we present our cases, and in Section 8.6 we analyse how the paradox is apparent in these cases. In the final section, we answer the question of the extent to which we can actually speak of a paradox and strategies that can deal with this paradox effectively.

8.2 Conditions for Internally Successful Pilot Projects

The internal success of a pilot project has to do with the extent to which the pilot successfully realises its main ambition. Often, this ambition relates to testing innovative approaches or concepts and developing new insights and knowledge. Subsequently, success can be considered as 'evidence' developed to inform policy decisions (Nutley, Davies and Smith, 2000; Nutley, Davies and Walter, 2002). Furthermore, the internal success of a pilot can be understood in terms of collaboration, as in the establishment of new/positive cooperation between stakeholders (Vreugdenhil, Taljaard and Slinger, 2012; Vinke-de Kruijf, Kuks and Augustijn, 2015; Van Popering-Verkerk and van Buuren, 2017).

Despite the unique context in which a pilot takes place, several conditions for a pilot's internal success can be found in the literature, depending on the pilot's objective. We have identified five, as follows.

8.2.1 Being a Safe Haven at a Distance

A first condition for a successful pilot is its status as a safe haven, at a distance from the day-to-day routines and established practices of the policy regime. This distance is important for a pilot to generate some room for experimentation, typically not present in the mother organisations themselves (Hargadon and

Douglas, 2001). Political support is often a crucial element to safeguard this condition and to prevent interventions by actors within the regime who try to influence the exploratory process of the pilot. In the literature, this condition is often labelled as a protective space (Smith and Raven, 2012; Smith et al., 2014; Fatimah, Raven and Arora, 2015).

8.2.2 Sufficient Resources for Exploration

A second crucial condition is the availability of sufficient resources to fulfil the pilot project. These resources include expertise and means to generate knowledge, time available for undertaking experiments, monitoring and analysis and attention and manpower from the participating public organisations (Sanderson, 2002; Vinke-de Kruijf et al., 2015).

8.2.3 Willing and Competent Boundary Spanners as Participants

Pilots are only successful if the actors involved are open to new perspectives and willing to learn. Such a coalition of the willing is an important success condition for pilot projects. Innovative processes often attract more entrepreneurial persons, or persons who are more able to span boundaries between different perspectives and ways of knowing (Williams, 2002; Hoppe, 2009; Kemp and Rotmans, 2009). Pilots need these boundary spanners.

8.2.4 Principled Engagement and Social Learning

Pilots are often collaborative efforts, especially when they address complex issues. For successful collaborative innovation, it is essential that actors are able to achieve frame reflection (by which actors reflect on their normative beliefs and problem perceptions) and social learning, via a process of principled engagement (Sørensen and Torfing, 2012; Vinke-de Kruijf et al., 2015). A process design that facilitates inclusive, open dialogue is conditional for such a learning process (Schusler, Decker and Pfeffer, 2003; Muro and Jeffrey, 2008, 2012).

8.2.5 Limited Scale

Often, a limited scale is a decisive factor for a pilot to be implemented. A limited scale reduces the impact and, therefore, the risks of the pilot project and allows for precise outcomes by setting clear boundaries in space and time (Gearheart, Klopp and Allen, 1989; Kylefors, Andreas and Lagerkvist, 2003; Knudsen et al., 2009). Limited scale is a relative notion depending on policy intentions. Limited can refer, for example, to a municipal level if the intention is a national rollout, or to a single floodplain if the idea is to be implemented at a river branch.

It is important to note that several of the conditions previously mentioned can be strengthened during the course of a pilot process. For example, a pilot that gains status is often able to attract additional resources for further exploration, or social learning can be reinforced by the social capital developed during the course of the exploration.

8.3 Conditions for Learning from Pilots and Uptake (External Success)

The wider impact of a pilot can be seen as its external success. We define this success as the ways in which the output of a pilot results into more enduring change within the policy regime (by leading to changes in policies, visions, rules and standards and so on) and whether its findings or results are also applied in other contexts (e.g. a larger area, time span or set of problems). This also holds if the pilot demonstrates that a certain solution does not work or only works under certain conditions. The pilot is then considered successful, if policies cancel using the innovation and if the solution is not being implemented elsewhere. The successful uptake of pilot results depends on a variety of factors, summarised in the following sections.

8.3.1 Normative Congruence: Sufficient Proximity between Pilot and Parent Organizations

For a successful uptake of pilot results, it is important that pilot participants keep in mind the dominant values in the policy regime and in their home organisations. Innovative ideas are more likely to succeed when they correspond sufficiently with existing insights and knowledge of actors, and in particular of the dominant actors (Hargadon and Douglas, 2001; Coenen et al., 2010; Eshuis and van Buuren, 2014). Too much distance in terms of differences between the shared beliefs developed within the pilot and the fundamental values and beliefs of the regime hinders the uptake of innovations (Hargadon and Douglas, 2001). Hence, the ideas should be connected to the existing understandings and actions and yet should be set apart. One step further is not only that the ideas fit the organisation but also that (some) participants act as ambassadors and boundary spanners: they stand by the innovation, are convinced that it represents a large improvement compared to current practice, actively promote it and bridge the different worlds of the pilot context and the home organisation (Boer and During, 2001Howells, 2006).

8.3.2 *Congruence of Pilot Outcomes with Standard Operating Procedures*

Secondly, it is important that there is a sufficient degree of compatibility of a pilot's outcomes with the standard procedures for resource allocation (Nair and Howlett, Chapter 9; Hargadon and Douglas, 2001). Public organisations are usually dominated by a new public management logic in which efficiency and risk minimization are crucial. Pilot results have at least to fit into this dominant set of rules and the distribution of resources to become acceptable for wider application in the ongoing operation.

8.3.3 *Representativeness of Pilot Team Composition*

For the wider uptake of pilot results, it is important that future users are involved in the experiment and that room is created for the heterogeneity of user demands (Franke and Von Hippel, 2003; de Moor et al., 2010). This holds even more strongly when the innovation is not user driven but supplier driven. Sufficient interaction between the niche and the mother organisations is necessary to ensure that the regime will value the pilot as a legitimate source of knowledge (Ingram et al., 2015). Such a strategy can be deployed by engaging representatives of the various home bases in the learning process in the pilot project (de Moor et al., 2010; Smink et al., 2015).

8.3.4 *Pilot as Instrument for Policy Change and Bandwagon*

Related to the second issue is the issue of embedding the pilots into wider environments (Hoogma, 2002; van den Bosch, 2010; Nair and Howlett, 2015). An important component of this embedding has to do with ensuring that the outcomes of a pilot can be easily linked to present problems and agendas of the receiving organisations (Hargadon and Douglas, 2001). Another important component has to do with paying attention to the use of the pilot outcomes during the pilot project: how a pilot can become an instrument of policy change (Nair and Howlett, Chapter 9).

8.3.5 *Perceived Added Value and Trust in Results*

In the *perception* of agents who have to adopt the results, the results of the pilot should be relevant, reliable, representative and useful (de Moor et al., 2010). This imposes quality criteria on the pilot project. Did the design fit the questions that had to be addressed, were the (temporal and spatial) scales appropriate to answer the questions, and were the 'right' questions addressed? A pilot can, for example, be easily set aside by classifying results as non-representative. For a further uptake of the pilot, the pilot results have to be trusted and considered to contribute sufficiently to formal policy objectives. Continuous participation by potential users may prevent this pitfall (Lettl, 2007; de Moor et al., 2010).

8.4 The Pilot Paradox

Within the world of international aid and development, 'pilots never fail, they (also) never scale' is a common expression (Dagerskog et al., 2016). Although pilots often succeed in testing innovative solutions, producing new knowledge and contributing to participant learning, they are commonly far less successful in generating enduring change. A successful pilot (which delivers useful knowledge and results in terms of more knowledgeable participants) is anything but a recipe for successful diffusion and upscaling.

We argue that, partly, the explanation for this has to do with the contradictory conditions for internal and external success. In other words, the conditions that contribute to a successful pilot are also major barriers to a successful uptake. With uptake we mean embedding and implementing the results of the pilot within the current policy regime. In Table 8.1, we summarise the two sides of the coin.

8.5 Case Studies

We explore the paradoxical relation between success conditions of pilot projects in six case studies. We used three main criteria for our case selection:

- Projects initiated as pilot project and running for several years
- Relevance to main policy issues regarding climate change adaptation in the Netherlands (flood risk management, fresh water availability, soil subsidence, ground water management, water nuisance)
- Data availability to allow for an analysis of internal and external success, and related conditions

The main characteristics of the selected case studies are summarised in Table 8.2.

The case studies are based on empirical material collected through a mixture of interviews, document analysis, focus groups and other methods (see Table 8.3). The case analysis and the comparative analysis were performed by all authors involved via collaborative interpretation and ranking.

In the following paragraphs, we describe these pilot projects, paying attention to the aim of the pilot project, the way the pilot project is/was organised and its results.

Table 8.1. *The pilot paradox*

Element	Conditions for successful pilots	Conditions for uptake
Position of the pilot	Freedom to explore novel ideas: degrees of independence (with regard to content, way of working, rules) of the principal organisations and flexible application of rules.	Keeping connected: conscious strategy to create normative congruence. Reporting to own organisation on different levels (political, strategic, tactical and operational).
Resource distribution	Extra resources (budget, expertise, time) for the pilot to enable creativity and exploration. Political attention/pressure to 'score' with the pilot.	Solutions fit within the existing system of resource distribution and contribute to organisational aims of efficiency and risk reduction.
Participants	Coalition of (willing) boundary spanners. Easy communication and openness about interests, etc. Participants are willing/able to experiment and take more vulnerable positions.	Representativeness of involved actors from all relevant disciplines and stakes of the future implementation arena, including users. Potential criticasters from participating organisations are included in the process, and participants act as ambassadors.
Process design	Learning environment, tailor-made collaborative process design. Freedom to organise own process and to get rid of traditional role distribution.	Results ready for mainstreaming and broader embedding. Potential future application areas are identified, and outcomes are linked to actual policy questions.
Project design	Limited scale to reduce risks and (financial) impacts, high-quality (shared) monitoring and analysis.	Sufficient system understanding; outcomes are trusted, considered as representative and relevant.

8.5.1 *Beuningen: River Maintenance Based on Cyclic Floodplain Rejuvenation*

In the floodplains of Beuningen, Rijkswaterstaat (RWS), the implementation agency of the Dutch Ministry of Infrastructure and Environment had identified increased water levels due to vegetation growth. Hence, the discharge capacity of the river no longer met the norms. Instead of simply removing all the vegetation, a nature organisation (Ark), together with Radboud University, developed a new concept called Cyclic Floodplain Rejuvenation (CFR). Free-flowing rivers regularly reset vegetation and lower sandbanks as a result of erosion and so make space for the water to flow. The idea of the concept is to imitate this natural behaviour and to create space, on the one hand, and increase natural amenity resulting from

Table 8.2. *Summary of the studied pilot projects*

Pilot	Start and end	Type of climate adaptation	Innovation	Level
Beuningen	2008–2010	Nature development and water level management	Technical	Local
Koopmanspolder	2011–ongoing	Nature development and water level management	Technical	Local
Landbouw op peil	2011–2014	Water level management and fresh water supply	Technical and process	Regional
Loosdrechtse plassen	2011–2013	Water level management	Technical	Local
Marken	2012–2015	Flood safety	Technical and process	Local
Water farming Walcheren	2010–ongoing	Fresh water supply	Technical and process	Local

Table 8.3. *Empirical material per case*

Case	Data collection
Beuningen (2008–2010)	Interviews (11), document analysis, observation workshops (Peters, Kater and Geerling, 2006; Vreugdenhil et al., 2010; Freude am Fluss, 2007)
Koopmanspolder (2011–ongoing)	Document analysis and participatory observation (https://publicwiki.deltares.nl/display/CAW/Koopmanspolder)
Landbouw op peil (2011–2014)	Document analysis, (group) interviews (35) (Kuindersma and Breman, 2014)
Loosdrechtse plassen (2011–2013)	Document analysis, observations, interviews (Ellen and Ottow, 2012; Breman et al., 2014)
Marken (2012–2015)	Document analysis, observations, interviews (van Buuren et al., 2015b)
Water farming Walcheren (2010–ongoing)	Document analysis, interviews (5) (van Buuren et al., 2015a)

larger variety, on the other. In 2004, the idea was launched to design the floodplain according to these principles. The idea was trialled far away from the RWS home base and resources were limited, particularly for implementation. Hence, it took until 2008 for the pilot to be implemented. The pilot was considered positive by the ambassadors, but, at the same time, it could be easily negated due to a lack of monitoring. Plans were developed to extend the pilot. The Ministry of the

Environment would finance this to create a climate buffer. However, RWS meanwhile reached its flood defence objectives and decided to follow the strategy of controlling and maintaining the situation as it is, meaning that there was no juridical space for the CFR approach in which vegetation develops more variety and that the external success of the pilot was limited.

8.5.2 Loosdrecht: Flexible Water Levels

In the recreational area of Loosdrecht, in the western part of the Netherlands, the water board decided to apply flexible surface water level management. This should allow the water level to rise and fall with precipitation and evaporation within certain preset boundaries. This is against traditional water management in the Netherlands, where water levels are set on a single level and do not fluctuate. The reason for doing so was to reduce the need to let in water from the surrounding waterways with high nutritional values in summer, which is harmful for the ecological goals set for the Loosdrecht area under the European Union Water Framework Directive (WFD). However, the new water regime caused several worries among citizens about the possible effects, including flooding and impediment to water recreation.

Because the pilot was a subsidised project – within the Dutch innovation programme of the WFD – the pilot was organised as an inter-organisational project with many external research institutes involved in the team. There was ample funding for both the actual implementation of the pilot and the monitoring and insights into the pilot's effects. The position of the pilot was well embedded within the organisation of the water board but only in the ecology-oriented part of the organisation. The process was designed around implementation of the pilot and the monitoring and dissemination of its results. Stakeholders including a municipality, marinas and house owners were mainly involved by means of participatory monitoring. The results of the pilot were framed as a success by the water board. The water board mainstreamed the concept of flexible surface water level management, thanks to the active dissemination of the pilot's results.

8.5.3 Koopmanspolder: A Paradise for Water Birds

The Netherlands needs new types of water management to be prepared for the impact of climate change. To create different functions that support nature, fish, and leisure facilities, the 'inland shore' concept, a new approach to water storage, was developed. A pilot was installed in the Koopmanspolder in the northwest of the Netherlands. The first tests in the Koopmanspolder are bearing fruit: the polder is partially submerged and nature is flourishing. The pilot was organised close to the home base of the primary coordinators of the project (the water board and RWS). Furthermore, the project consortium consisted of the province, the municipality, Deltares, Staatsbosbeheer and a number of local non-governmental organisations (NGOs). Funding for the pilot came from different sources, but the pilot was primarily funded by the National Fund for Rural Areas (ILG). The pilot has a very long history, which started in the 1990s when the area was made part of the ecological structure of the Netherlands, but, over time, the focus changed from nature development to more multifunctional purposes. This resulted in an award-winning landscape design for the polder in 2011 that combined nature development with flood-risk management, recreation and fresh water supply. The pilot was considered a success both within and outside the coordinating organisation. This resulted in the start of a second pilot of the inland shore concept by public and private parties, with a focus on agriculture, aquaculture and fisheries in other parts of the province.

8.5.4 Marken: A New Approach to Flood Safety

The water safety situation on the island of Marken in the Netherlands is not adequate to resist high water levels in the future. The dikes around the island were rejected in former test rounds. Therefore, the National Water Agency developed a dike reinforcement plan in 2012. This plan did not win the support of the island's inhabitants because of the impact on cultural heritage and the adjacent houses. Furthermore, the plan was too expensive. At the same time, the Netherlands was developing a new water safety strategy that paid more attention to spatial planning and evacuation strategies in the event of floods. For this reason, the involved governments developed a pilot to investigate whether the new water safety strategy would be a solution for the island of Marken (a technical innovation). The pilot was a cooperation between the National Water Agency, the municipality, the water board and local stakeholders. The involved stakeholders worked together in a project group that functioned as a safe haven for the project workers. The pilot was governed by a steering committee and funded by the national government. The time horizon for the pilot was three years. The Marken case was successful in terms of developing innovative water safety strategies for the island (smart development of dikes) based on new water safety principles. The results of the pilot will be used for other vulnerable water safety situations in other regions in the Netherlands.

8.5.5 Landbouw op Peil

Climate change poses new challenges for Dutch water managers. This also holds for the water boards in the Rijn-Oost region. Most of the area that these water boards manage consists of agricultural lands. Therefore, one of the biggest challenges for these water boards is to develop a future-proof system of water management that leaves sufficient room for vital agricultural production systems. This requires innovative measures optimising a sustainable water system in relation to the agricultural function. With this challenge in mind, the Landbouw op peil (Agriculture at Water Level) project was started in 2011. The aims of the project were multi-faceted:

- To gather knowledge and experience on how to maintain a vital agriculture within a changing system of water management (due to climate change)
- To improve the relation between the agricultural sector and water managers
- To contribute to a climate-proof agriculture in the eastern part of the Netherlands
- To use local knowledge from agricultural firms to improve future water management

The pilot is perceived as very successful by almost all of the participating stakeholders. One of the main factors for this internal success is the feeling amongst stakeholders that the pilot has really contributed to increased levels of trust and understanding, which has helped to narrow the gap between stakeholders. In addition, most of the measures designed in the pilot for dealing with the water challenges were accepted by the farmers and were continued after the pilot had finished. The pilot received quite a lot of (media) exposure and was framed as (very) successful, both internally and externally. At the end of the pilot stage, the pilot won an award (for innovative projects) accompanied by a sum of money. This money was reinvested in a follow-up to the pilot enabling the possibility of drawing up more water deployment plans (*waterbedrijfsplannen*) for other interested farmers. These tailor-made plans at farm level are considered to be one of the crucial success factors. This has contributed to a (moderate) external success for the pilot as well.

8.5.6 Water Farming Walcheren: A Pilot to Improve Fresh Water Availability for Local Farmers

Walcheren is a region in the Netherlands where agriculture is an important economic activity. Increasingly, the consequences of climate change are being felt (shortage of fresh water supply). This negatively influences farmers' yields. Farmers, therefore, started to cooperate to improve the availability of fresh water in dry periods by conserving the water in the soil and watercourses. They started a pilot to test the technical possibilities to infiltrate the water into the soil in winter and increase the water levels in watercourses. The pilot is a private initiative of farmers facilitated by knowledge institutes. The water board provides permits but its involvement is mostly from a distance. Therefore, the pilot is rather far away from the home bases of the water authorities. The pilot has few resources available (only private investments and subvention from public parties). The pilot process can be characterised as 'learning by doing'. No end date is stated for the pilot. Based on their effectiveness, technical measures are added in the pilot. The pilot is successful in the effectiveness of chosen measures, but less successful in generating enduring policy change. The pilot is locally customised and based on the physical characteristics of the terrain.

8.6 Analysis of the Paradox

Table 8.4 presents a summary of our analysis of how the pilot projects satisfy the success conditions proposed in Sections 8.2 and 8.3. For each case, we indicate whether this condition was present in the pilot:

- Condition was absent in the pilot (double minus)
- Condition was almost absent in the pilot (minus)
- Condition was partly present in the pilot (zero)
- Condition was present in the pilot (plus)
- Condition was dominant in the pilot (double plus)

We assessed the presence/absence by coding our empirical material and looking for the extent to which the various indicators per condition could be found. Dependent on the question of how many indicators we found as well as how strongly their presence was valued by our respondents, we decided how to assess the various conditions. We discussed the scores several times in the team of authors and used the comparison of the various cases to fine-tune the individual scores. In the sections following Table 8.4 and Table 8.5 we describe the empirical observations our scores are derived from.

8.6.1 Explanation of the Scores

Beuningen

The pilot was initiated by a nature organisation in cooperation with a university. Resources to develop the concept and a site were available, but one of the main issues to be addressed was who should pay for the pilot. The innovation fell between existing administrative structures: the question arose as to whether it was a measure for nature development or for flood protection. These interests were covered by two different ministries. Hence, the discussion arose on who was financially responsible. This delayed the pilot. In combination with new model results that reduced the actual

Table 8.4. *Case comparison*

	Conditions for internal success					Conditions for external success				
	Distance from home base	Additional resources	Coalition of boundary spanners	Tailor-made process design	Low-risk scale	Normative congruence	Organisational appropriateness	External representativeness	Pilot embedding	System understanding
Beuningen	+	+	+	+	+	-	-	+	0	0
Koopmanspolder	+	++	++	+	+	+	+	++	+	0
Landbouw op peil	+	++	+	++	+	+	-	0	-	+
Loosdrecht	+	++	-	++	+	+	+	-	+	+
Marken	+	++	++	0	++	++	0	-	-	+
Water farming Walcheren	++	+	+	++	++	-	-	+	-	0

Table 8.5. *Success evaluation of the pilots*

	Results of the pilot projects	
	Internal success: Results of the pilot project	External success: Mainstreaming of results
Beuningen	Moderate: little monitoring, pilot already 'died' before implementation	Low: little fit with existing institutional structures and culture
Koopmanspolder	Very high: attempts are being made to replicate the pilot	High: other locations are also being considered, using the same consortium
Landbouw op peil	Very high: all stakeholders satisfied; increased trust and improved relations	Moderate: there was little infrastructure to deal with the increased interest amongst farmers; no structural follow-up
Loosdrecht	Moderate: the pilot was considered a success but has a distinct user group	Moderate: the measure is being considered by other public authorities
Marken	High: successful development of customised water safety strategy	Low: newly developed water safety strategy is only slightly embedded in daily routines
Water farming Walcheren	Very high: stakeholders typify their pilot as successful	Low: solutions do not automatically fit in the policy system

need to take action, the pilot was downsized significantly, both in size and monitoring efforts. In combination with a second issue – misfit with institutions and their culture – little scaling-up has been achieved. The idea of using river dynamics does not fit with how the river authorities work.

Koopmanspolder

This pilot took a very long time to find its final shape – starting in the 1990s and developing into the pilot process that was initiated in 2011 and continuing to this day. The available resources brought together by the coalition of the willing led to its success, and it was actually also this coalition that formed the basis for its successful uptake (other locations for piloting are being considered). However, one can ask whether the concept will progress beyond a string of pilots and be mainstreamed when the concept has to leave the safe haven of the pilot context.

Landbouw op Peil

One of the main factors for the (internal) success of the Landbouw op peil project was the room for tailor-made solutions at individual farm levels. In a process of intensive guidance, specific water management plans were discussed and developed for each participating farm, thus targeting the specific (water) challenges at farm level with tailor-made measures and solutions. This approach resulted in a high level of enthusiasm, commitment and confidence among the participants. To be able to implement these tailor-made solutions, the water managers also left quite some room for manoeuvre and sometimes distanced themselves from the existing practice and protocols within their organisations. Within the pilot, this resulted in a very productive and fruitful dynamic. However, as this dynamic was also quite far away from common practice, it hindered the up-scaling/follow-up of the pilot.

Loosdrecht: Flexible Water Levels

The Loosdrecht pilot meets almost all the conditions for a successful pilot – the only thing lacking is boundary spanners, as the pilot was more designed top-down and implemented by the water board – overriding the protests of the stakeholders in the area. Furthermore, the pilot clearly established a normative congruence with its home organisation, and, although the stakeholders protested, the pilot also made it possible to involve them as participants.

Marken

The Marken case scores high on the conditions for successful pilots. It was organised at a safe distance from normal routines. Furthermore, there was a coalition of boundary spanners. The pilot had sufficient additional resources. The case scores relatively low on the conditions for successful mainstreaming. The water safety strategy developed was highly locally customised and only slightly embedded in the daily routines of the involved organisations. Involved stakeholders told us that the approach in this pilot was too time-consuming to make it common practice.

Water Farming

The Water Farming Walcheren case is another typical illustration of the pilot paradox. It is organised at a far distance from water authorities and has a tailor-made process design. Its weakness, however, is its distance from the dominant values in the water authorities' home bases. The pilot's embedding and organisational appropriateness score low. At the same time, there is a lack of normative congruence between the stakeholders in the pilot and the policy regime. This hinders mainstreaming the pilot results.

8.6.2 Interpretation of Results

The first observation is that, in the majority of the cases (four of the six cases), we can indeed witness that a high score on the conditions for internal success is associated with a low score on the conditions for wider uptake (Marken, Water Farming, Beuningen and Landbouw op peil). In other words, there is indicative evidence of a pilot paradox in the cases analysed. At least two mechanisms can explain this.

Firstly, there is the mechanism of *goal displacement*. Actors within a pilot do a really good job of making the pilot successful, because of the attention and time they give to it during the pilot, but they fail to keep the wider environment connected. They focus on realising the objectives of the pilot in producing an innovative outcome, which they manage by mobilising additional resources. All available capacity is deployed to make the pilot itself successful, thereby decreasing attention on the broader dissemination of the outcomes in policies and practices. Of course, this is also much less energising than developing the innovation itself. However, because of this goal displacement, the pilot project loses relevance. The pilot becomes a goal in itself, and this results in myopia: uptake is overlooked.

The second, related and complementary mechanism can be called *regime alienation*, or growing myopia among the pilot participants. This mechanism was found in at least three projects and was clearly visible, for example, in the Marken pilot. As a result of intensive learning processes at a distance from the home bases, the distance to the mother organisations becomes greater instead of smaller. In that situation, the outcomes of the pilot are so different from what is acceptable for the regime that wider uptake becomes increasingly difficult. In the Marken pilot for example, it proved almost impossible to transplant pilot results into other flood protection projects, which are traditionally focused on dike enforcement.

At the same time, two of the six pilots show far fewer elements of the paradox or even show good results both internally and externally (i.e. Koopmanspolder and Loosdrecht), although uptake of their results is also mainly conceptual and relates to policies rather than substance and implementation. We can see at least two explanations for simultaneous internal and external success.

First of all, in some cases, the pilot results come at the right moment. There is – in other words – a *window of opportunity* for the outcomes of the pilot to become mainstreamed in the policies and practices of the various home bases (see the Loosdrecht case). Frequently, a long incubation phase before the start of the pilot can result in a fertile ground for its results. In such cases, the pilot is the result of growing regime frustration because of the ineffectiveness of current solutions. The mother organisations are actually waiting for the results of the pilot because they are not able to solve the problems in the traditional way and they have to arrive at alternative solutions.

Secondly, in a couple of cases, the pilot paradox is effectively prevented by *self-correcting mechanisms* within the pilot process, for example, by making the broader uptake of the lessons of the pilot explicitly part of the aims of the pilot or by the way the involvement of actors is organised. Self-correcting mechanisms are visible in those pilots that are not very well embedded in their wider environment or whose representativeness is low. These pilots result in rather innovative results that are also congruent with the dominant beliefs in the wider policy environment (Loosdrecht). In the Koopmanspolder case, the pilot is organised in such a way that participants continuously inform their environment about what is going on and the preliminary results. This case is also the only case in which all conditions for internal and external success were met. In this case, we can witness a carefully crafted strategy to organise an intelligent process for exploration in a safe environment and an intense process of dissemination to safeguard the wider uptake of the results. In other words, a rather ambidextrous strategy was used to develop the innovation and to organise its uptake.

8.7 Conclusions: The Paradox Revisited

From this analysis, we can draw a couple of conclusions. First of all, we can conclude that often there is a paradoxical relationship between the conditions for the internal and the external success of pilot projects. Focusing on meeting the conditions for internal success creates a barrier against external success. In particular, the distance between pilots and the regime, and the dedicated process to find an innovative solution, forms a barrier to arriving at appropriate outcomes and organising sufficient embedding necessary for wider uptake in the existing regime.

At the same time, the pilot paradox is far from a general causal mechanism. It is possible to prevent it, especially by carefully designing the process or course of events in the pilot and by formulating its aims in terms of both internal and external success. Making broader uptake part of the main targets and focusing attention on this target from scratch are not only very logical but also rather effective ways to overcome the paradox.

This brings us to the (managerial) implications of the paradox. Designing experiments with impact is only possible when actors recognise the pilot paradox with which they can be confronted. Within the pilot management design, attention must be paid to issues of external representativeness and boundary spanning between the niche and the regime (by way of dialogue with the wider environment). This can be organised, for example, by involving future users in an early phase of the pilot, or by starting to communicate some of the (intermediate) successes and obstacles to a wider public from early on in the pilot. The distance from the existing regime is something else that has to be carefully balanced: organising enough distance to be free to improvise is indispensable, but the same holds true for organising provisions to safeguard embedding. And finally, it is absolutely necessary to organise learning and reflection not only after finalizing the pilot but also during the execution of the pilot to enable timely redirection and adjustment of the course of the pilot and to prevent goal displacement.

References

Banks, G. (February 2009). Evidence-based policy making: What is it? How do we get it? How do we get it? ANU Public Lecture Series, presented by ANZSOG, Canberra.

Berkhout, F., Verbong, G., Wieczorek, A. J., Raven, R., Lebel, L., and Bai, X. (2010). Sustainability experiments in Asia: Innovations shaping alternative development pathways? *Environmental Science & Policy*, 13(4), 261–271.

Boer, H., and During, W. E. (2001). Innovation, what innovation? A comparison between product, process and organisational innovation. *International Journal of Technology Management*, 22(1–3), 83–107.

Bos, J. J., Brown, R. R., and Farrelly, M. A. (2013). A design framework for creating social learning situations. *Global Environmental Change*, 23(2), 398–412.

Breman, B. C., de Groot, M., Ottow, B., and Rip, W. (2014). Monitoren doe je samen. De meerwaarde van participatieve monitoring. *H2O Online*, 20 (6), 1–9.

Coenen, L., Raven, R., and Verbong, G. (2010). Local niche experimentation in energy transitions: A theoretical and empirical exploration of proximity advantages and disadvantages. *Technology in Society*, 32(4), 295–302.

Dagerskog, L., Savadogo, K., Hamadou, K., and Vodounhessi, A. (2016). Productive sanitation in Burkina Faso and Niger – going beyond projects? Paper presented at the 5th International Dry Toilet Conference, Tampere, Finland, 19 August 2015–22 August 2015.

De Moor, K., Berte, K., De Marez, L., Joseph, W., Deryckere, T., and Martens, L. (2010). User-driven innovation? Challenges of user involvement in future technology analysis. *Science and Public Policy*, 37(1), 51–61.

Ellen, G. J., and Ottow, B. T. (2012). *Maatschappelijke aspecten van flexibel peilbeheer*. Utrecht: Deltares.

Eshuis, J., and van Buuren, A. (2014). Innovations in water governance: The importance of time. *International Review of Administrative Sciences*, 80(2), 401–420.

Fatimah, Y. A., Raven, R. P., and Arora, S. (2015). Scripts in transition: Protective spaces of Indonesian biofuel villages. *Technological Forecasting and Social Change*, 99, 1–13.

Franke, N., and Von Hippel, E. (2003). Satisfying heterogeneous user needs via innovation toolkits: The case of Apache security software. *Research Policy*, 32(7), 1199–1215.

Freude am Fluss (2007). *Freude am Fluss. An Innovative Approach to River Management*. Beuningen, the Netherlands: Drukkerij Libertas.

Gearheart, R. A., Klopp, F., and Allen, G. (1989). Constructed free surface wetlands to treat and receive wastewater: Pilot project to full scale. In Hammer, D. A. (ed.), *Constructed Wetlands for Wastewater Treatment*. Chelsea, MI: Lewis Publishers, 121–138.

Hargadon, A. B., and Douglas, Y. (2001). When innovations meet institutions: Edison and the design of the electric light. *Administrative Science Quarterly*, 46(3), 476–501.

Hoogma, R., Kemp, R., Schot, J., and Truffer, B. (2002). *Experimenting for Sustainable Transport: The Approach of Strategic Niche Management*. London: Spon Press.

Hoppe, R. (2009). Scientific advice and public policy: Expert advisers' and policymakers' discourses on boundary work. *Poiesis & Praxis*, 6(3–4), 235–263.

Howells, J. (2006). Intermediation and the role of intermediaries in innovation. *Research Policy*, 35(5), 715–728.

Ingram, J., Maye, D., Kirwan, J., Curry, N., and Kubinakova, K. (2015). Interactions between niche and regime: An analysis of learning and innovation networks for sustainable agriculture across Europe. *Journal of Agricultural Education and Extension*, 21(1), 55–71.

Kemp, R., and Rotmans, J. (2009). Transitioning policy: Co-production of a new strategic framework for energy innovation policy in the Netherlands. *Policy Sciences*, 42(4), 303–322.

Kivimaa, P., Hildén, M., Huitema, D., Jordan, A., and Newig, J. (2017). Experiments in climate governance: A systematic review of research on energy and built environment transitions. *Journal of Cleaner Production*, 169, 17–29.

Knudsen, J. N., Jensen, J. N., Vilhelmsen, P. J., and Biede, O. (2009). Experience with CO_2 capture from coal flue gas in pilot-scale: Testing of different amine solvents. *Energy Procedia*, 1(1), 783–790.

Kylefors, K., Andreas, L., and Lagerkvist, A. (2003). A comparison of small-scale, pilot-scale and large-scale tests for predicting leaching behaviour of landfilled wastes. *Waste Management*, 23(1), 45–59.

Kuindersma, W., and Breman, B. C. (2014). Leren van landbouw op peil. Evaluatie van een experiment met zelfsturing in het waterbeheer. Alterra report 2512.

Lettl, C. (2007). User involvement competence for radical innovation. *Journal of Engineering and Technology Management*, 24(1), 53–75.

May, P. J., and Jochim, A. E. (2013). Policy regime perspectives: Policies, politics, and governing. *Policy Studies Journal*, 41(3), 426–452.

Muro, M., and Jeffrey, P. (2008). A critical review of the theory and application of social learning in participatory natural resource management processes. *Journal of Environmental Planning and Management*, 51(3), 325–344.

Muro, M., and Jeffrey, P. (2012). Time to talk? How the structure of dialog processes shapes stakeholder learning in participatory water resources management. *Ecology and Society*, 17(1), 3.

Nair, S., and Howlett, M. (2015). Scaling up of policy experiments and pilots: A qualitative comparative analysis and lessons for the water sector. *Water Resources Management*, **29**(14), 4945–4961. doi: 10.1007/s11269-015-1081-0.

Nutley, S., Davies, H., and Walter, I. (2002). Evidence based policy and practice: Cross sector lessons from the UK. Working Paper, 9. London: ESRC UK Centre for Evidence Based Policy and Practice, University of London.

Nutley, S. M., Davies, H. T., and Smith, P. C. (2000). *What Works? Evidence-Based Policy and Practice in Public Services*. Bristol: Policy Press.

Peters, B. W. E., Kater, E., and Geerling, G. W. (2006). *Cyclisch beheer in uiterwaarden: Natuur en veiligheid in de praktijk*. Nijmegen: Centrum voor Water en Samenleving, Radboud Universiteit.

Rotmans, J., and Loorbach, D. (2009). Complexity and transition management. *Journal of Industrial Ecology*, 13(2), 184–196.

Sanderson, I. (2002). Evaluation, policy learning and evidence-based policy making. *Public Administration*, 80(1), 1–22.

Schusler, T. M., Decker, D. J., and Pfeffer, M. J. (2003). Social learning for collaborative natural resource management. *Society & Natural Resources*, 16(4), 309–326.

Smink, M., Negro, S. O., Niesten, E., and Hekkert, M. P. (2015). How mismatching institutional logics hinder niche–regime interaction and how boundary spanners intervene. *Technological Forecasting and Social Change*, 100, 225–237.

Smith, A., Kern, F., Raven, R., and Verhees, B. (2014). Spaces for sustainable innovation: Solar photovoltaic electricity in the UK. *Technological Forecasting and Social Change*, 81, 115–130.

Smith, A., and Raven, R. (2012). What is protective space? Reconsidering niches in transitions to sustainability. *Research Policy*, 41(6), 1025–1036.

Sørensen, E., and Torfing, J. (2012). Introduction: Collaborative innovation in the public sector. *Innovation Journal*, 17(1), 1–14.

Van Buuren, A., Keessen, A. M., Van Leeuwen, C., Eshuis, J., and Ellen, G. J. (2015a). Implementation arrangements for climate adaptation in the Netherlands: Characteristics and underlying mechanisms of adaptive governance. *Ecology and Society*, 20(4), 11.

Van Buuren, M. W., Ellen, G. J., van Leeuwen, C., and van Popering-Verkerk, J. (2015b). *Die het water deert die het water keer: overstromingsrisicobeheer als maatschappelijke gebiedsopgave, opbrengsten en lessen uit de pilots meerlaagsveiligheid*. Rotterdam: Erasmus University Rotterdam.

Van Buuren, M. W., and Loorbach, D. (2009). Policy innovation in isolation? Conditions for policy renewal by transition arenas and pilot projects. *Public Management Review*, 11(3), 375–392.

Van den Bosch, S. J. M. (2010). Transition experiments: Exploring societal changes towards sustainability. PhD thesis, Erasmus University Rotterdam.

Van Popering-Verkerk, J., and van Buuren, M.W. (2017). Developing collaborative capacity in pilot projects: Lessons from three Dutch flood risk management experiments. Journal of Cleaner Production, 169, 225–233.

Vinke-de Kruijf, J., Kuks, S. M. M., and Augustijn, D. C. M. (2015). Governance in support of integrated flood risk management? The case of Romania. *Journal of Environmental Development*, 16, 104–118.

Vreugdenhil, H., Slinger, J., Thissen, W., and Ker Rault, P. (2010). Pilot projects in water management. *Ecology and Society*, 15(3), 13.

Vreugdenhil, H., Taljaard, S., and Slinger, J. H. (2012). Pilot projects and their diffusion: A case study of integrated coastal management in South Africa. *International Journal of Sustainable Development*, 15(1–2), 148–172.

Vreugdenhil, H. S. I. (2010). Pilot projects in water management: Practicing change and changing practice. PhD thesis, Delft University of Technology.

Williams, P. (2002). The competent boundary spanner. *Public Administration*, 80(1), 103–124.

Policy Pilots for Climate Adaptation in Indian Agriculture

A Qualitative Comparative Analysis

SREEJA NAIR AND MICHAEL HOWLETT

9.1 Introduction to Policy Piloting as a Form of Governance Experimentation

Climate change is a policy problem characterized by a high degree of uncertainty (Parry et al., 2007). Environmental degradation and change can lead to certain policy thresholds being crossed, limiting the effectiveness of current policy responses in the long-term (Kwadijk et al., 2010). And for long-term policies, such as those dealing with environmental issues, policymakers grapple with uncertainties in the policy formulation stage owing to a lack of complete understanding of the biophysical and social systems affecting and being affected by environmental changes. These uncertainties may consequently lead to over- or underestimation of policy problems and thereby the articulation and enactment of 'solutions' that are ineffective or even counterproductive (Deyle, 1994). Piloting with new and alternative strategies or changes, hence, is often thought to prove useful under such conditions of risk and uncertainty (Swanson and Bhadwal, 2009), allowing major government policies and programmes to be pretested before launching these fully, and at a wider scale (Cabinet Office, 2003).

In the context of climate change and of increasing need for policy innovation, pilots and experiments can provide a proof of principle for some of the planned activities to help alleviate the governance gaps and deadlock in negotiations (Jordan and Huitema, 2014). Enhanced experimentation and learning are thought to be instrumental in 'keeping pace with the dynamic drivers and expressions of risk' (O'Brien et al., 2012). Different types of pilots can be designed. Some may involve new and substantial policy change while others might make only small adjustments to current policies and programmes (Majone, 1991).

Many pilots have been studied and documented in isolation, making them a context-specific subject, and generalizations of their learning effects have not been possible. Hoffmann (2011) argues that rather than such a piecemeal approach, it is more valuable to look at experiments collectively along with their combined impacts. This chapter focuses on policy pilots, understood as a particular form of governance experimentation initiated by national and sub-national governmental agencies, to enable adaptation to climate risks in the agriculture sector in India.

Since the beginning of India's first five-year development plan in 1951, the Government of India has established many pilots to address risks to agriculture production, especially in rainfed areas. The agriculture sector has high policy significance to the country's development and allows the study of pilots which encompass a wide range of variation in policy responses to deal with current and expected changes in the environment associated with climate change.

India has the largest land area under rainfed agriculture in the world, but ranks among the last in terms of yields from rainfed agriculture (GoI, 2011). The Government of India identifies six main risks to rainfed agriculture in India. These relate to (1) production (due to weather, pests, diseases, etc.); (2) price/market (input and output price volatility); (3) finance and credit (cash flow problems, limited access to credit and finance); (4) institutions (changes in regulations that influences farmers' activities); 5) technology (risk associated with new technology adoption); and (6) personal risk to life and assets of the farmer (GoI, 2007). Adaptation policy action in response to climate risks to agriculture can include the introduction of climate-hardy crops, infrastructural upgradation such as new irrigation systems, along with capacity building, education and training, innovative insurance schemes or changes in land management practices, among others (Christiansen, Olhoff and Traerup, 2011).

In this study, policy pilots are identified as a form of governance experimentation by government and government-affiliated agencies to help reduce production risks and uncertainties in the agriculture sector in India. Scaling up has been defined in different ways (Gillespie, 2004; Hartmann and Linn, 2007; Vreugdenhil, Taljaard and Slinger, 2012; see also Introduction and Conclusions). This study considers scaling up when more actors, policy components and administrative layers are added to the pilot, increasing its scope and thereby complexity (Vreugdenhil et al., 2012). Using a comparative case analysis of the design features of fourteen agriculture policy pilots in India, this chapter asks the question, do the design characteristics of policy pilots influence their scaling up and policy integration?

9.2 Policy Piloting and Development of New Policy Mixes

Policies typically form as 'mixes' of policy goals and means (to achieve these goals) through processes of policy change, with addition and subtraction of elements (goals and instruments) over time (Howlett and Rayner, 2013).

A specific policy mix comprises some abstract or conceptual goals, specific programme content or objectives and operational settings or calibrations (Hall, 1993; Cashore and Howlett, 2007; Howlett and Cashore, 2009). A key challenge while designing policies and policy pilots for the future is that they are launched to operate and interact in a space where there are pre-existing policy mixes that have developed over time (Howlett and Rayner, 2013).

Howlett and Rayner (2007) argue that the degree of coherency between policy goals and degree of consistency between policy means should be studied on a case-by-case basis. Typically, policy goals are considered to be coherent if they logically relate to the same overall policy aims and objectives and can be achieved simultaneously without any significant trade-offs. Policy goals are considered to be incoherent if they contradict the previous goal. Policy means are considered to be consistent when they complement each other and work in combination towards meeting a policy goal, and inconsistent when they work at cross-purposes (Kern and Howlett, 2009).

Hall's (1993) work on policy dynamics and policy change based on the three-order model remains the most quoted piece of literature on studying policy change. However, policy scholars in the last decade have argued that such aggregation can lead to a rather myopic view of the more complex and granular processes of policy change that may go beyond the incremental change vs. paradigmatic change classification (Howlett and Cashore, 2009).

Cashore and Howlett identify six elements that characterize a policy mix (Cashore and Howlett, 2007). This includes policy goals (general ideas that govern policy development), formal policy objectives and policy settings (on the ground requirements of the policy), instrument logic or norms guiding implementation preferences, mechanisms or types of instruments that are being utilized and calibrations or the specific ways in which the instrument is used.

The policy literature remains rather inconclusive on whether, under conditions of uncertainty, policymakers (in this case national and state-level government officials designing the pilots) prefer to pilot (test) incremental changes to existing policies or use it as an opportunity to innovate and undertake major policy changes. Experimental processes of policy formulation can allow for unexpected and unusual combinations of means- and ends-related components to occur and lead to mechanisms such as policy layering, conversion and drift that may either happen rapidly or gradually over time (Thelen, 2003; Beland, 2007). Furthermore, means-related changes may occur in the absence of corresponding shifts to policy aims, and conversely, ends may change without any alterations to the means of achieving them (Kern and Howlett, 2009). When such novel policy configurations emerge, these can often be unexpected, deviant from standard practice or even a suboptimal arrangement to address the policy problem (Wilder and Howlett, 2014).

Policy pilots are a special form of deliberate experimentation initiated by governmental agencies that can be instrumental for evaluating new programmes at a 'controlled small-scale' before introducing full-scale programmes (Swanson and Bhadwal, 2009; Weiss, 1975). Pilots are seen as useful in providing insights for dealing with complex policy issues featuring high levels of uncertainty (Vreugdenhil et al., 2012). When policy pilots are found to be effective, they may be 'diffused', i.e. continued or expanded. Diffusion of pilots can occur via replication of the pilot into other or similar pilots and 'scaling up' into more permanent policies or bigger pilots (Vreugdenhil et al., 2009).

There is a lack of empirical studies on the scaling up of pilots and their contribution to altering policy mixes (Vreugdenhil et al., 2012). While individual pilots have been studied, these are rather fragmented, and a comprehensive study on the empirical evidence on the composition of such pilots and the processes which lead to their diffusion, or not, is lacking. Furthermore, the study of the impact of pilots on policy development is often limited to 'learning from failure' (Vreugdenhil et al., 2009). Many factors influence these dynamics, including the pilot design and the context within which it is introduced. These can include factors such as the stakeholders involved, the availability of knowledge and resources behind the design process and outcomes, the choice of scale and the choice for pilot sites, the mode of governance that influences the nature of stakeholder engagement and learning, the level of innovativeness of the pilot and how it converges or diverges from the current policy context, flexibility to make changes to adapt to local conditions and, finally, the timing of the strategy for pilot diffusion (Vreugdenhil et al., 2009).

Any change in policy response brought about by policy pilots will typically be exposed to resistance from stakeholders, particularly those with vested interests (van Buuren et al., Chapter 8). This makes it difficult to introduce any radical changes in the policy mix even if new policy objectives are put forth (Kern and Howlett, 2009) and would attempt to fit through processes of 'learning, coercion and negotiation' (Rip and Kemp, 1998; Christiansen et al., 2011; Jordan and Huitema, 2014).

9.2.1 Case Description

Many schemes have been launched by the Government of India for the benefit of rainfed areas (GoI, 2007). While risk management strategies can be undertaken by farmers or farmer groups at the individual or community level informally, this chapter focuses on policy pilots that are publicly provided mechanisms launched by the Central Ministry of Agriculture, Government of India, towards management of risks and that enable adaptation in the sector. These can be categorized broadly under initiatives that cover agriculture extension, supply of quality seeds and inputs, pest management, infrastructure provision, social assistance especially during disasters, loan assistance and access to credit and insurance, relaxation in grain procurement procedures, supply of fodder and cash transfer (GoI, 2007).

The discussion in this chapter is based on cases of policy pilots implemented in India between 1990 and 2015, with 1990 marking the beginning of major reforms in the country, including the devolution of powers to the state-level[1] and economic liberalization in India, both impacting the agriculture sector. Consultations and interviews were conducted in 2014 and 2015 with members of the Working Group on Natural Resource Management and Rainfed Agriculture – which was created by the Planning Commission, Government of India – and other agriculture experts and members from international donor agencies working with the central and state governments in India on agriculture risk management to explore the landscape of agriculture policy pilots since 1990 followed by detailed information on individual cases.

The fourteen cases (see Table 9.1) include a mix of restructured policies as well as new pilots to address environmental (including climatic) risks to the agriculture sector. They represent models of pilot testing policy elements with the intention of guiding agriculture policy development at the national level.

The common features that the cases share include the following:

- The initiatives are aimed at addressing risks to agriculture production, including weather and climate-related risks in rainfed areas in the country, launched by the Central Government of India directly, in partnership with state government (s) and/or external funding.
- The cases are time-delimited, i.e. designed to run for a limited number of years and/or spatially limited with an explicitly stated intent (goal) of diffusion beyond the initial identified scope.
- The cases were identified as having been in a test phase aiming at guiding national policy development through testing of untried components.
- The cases have been subject to periodic monitoring and should have completed at least one round of formal monitoring and evaluation.
- These cases have finished the 'experimental' or pilot phase so that the outcome can be clearly studied.

9.2.2 *Qualitative Comparative Analysis*

A set-theoretical approach was found to be useful to study the relation between sets of change to a current policy mix (characteristics of the pilot design) and scaling up

---

[1] Agriculture is a state-subject in India.

Table 9.1. *Overview of the selected policy experiments*

No.	Name	Initial goal	Effective duration	Initial piloting unit	Scaling up outcome
1	Experimental Crop Insurance Scheme	Provide risk transfer mechanism for farmers	One season (1997/1998)	14 districts in 5 states	Terminated within 1 season
2	Farm Income Insurance Scheme	Provide risk transfer mechanism for farmers	One season (2003/2004)	15 districts in 8 states	Expanded to 19 districts but terminated within one season
3	Weather-Based Crop Insurance Scheme	Provide risk transfer mechanism for farmers	2007–2013	70 hoblis (cluster of villages) in Karnataka state	Merged with 2 pilots into a National Crop Insurance Programme
4	Modified National Agriculture Insurance Scheme	Provide risk transfer mechanism for farmers	2010–2013	32 districts in 12 states	Merged with 2 pilots into a National Crop Insurance Programme
5	Restructured National Watershed Development Project for Rainfed Areas	Integrate watershed management and sustainable farming	1990–1995	99 districts in 16 states	Continued to expand to additional states following revised watershed guidelines
6	Indo-German Watershed Development Programme	Regulate natural resources and soil and water conservation	1992–1999	1 district of Maharashtra	Created a Watershed Development Fund in 1991–1992 to replicate this model and sustain its activities; expanded to 4 additional states
7	Sujala Watershed Development Project	Increase agricultural productivity, and poverty alleviation	2001–2009	Watersheds in 5 drought-prone districts of Karnataka	Expanded to additional pilots and management programmes at the state level; guided national watershed policy development
8	National Project on Organic Farming	Improve soil health	2004–2014	State level	Subsumed under National Mission on Sustainable Agriculture
9	Rainfed Area Development Programme	Ensure agricultural growth in rainfed areas	2011–2014	District level in 10 states	Subsumed as a key component under National Mission on Sustainable Agriculture

171

Table 9.1. (*cont.*)

No.	Name	Initial goal	Effective duration	Initial piloting unit	Scaling up outcome
10	National Agriculture Innovation Project (NAIP)	Increase agricultural productivity and growth	2006–2014	Multiple projects launched country-wide	Gave rise to changes in institutional structure and function of the Indian Council for Agriculture Research, the research arm of the Ministry of Agriculture
11	National Initiative for Climate Resilient Agriculture	Ensure climate resilience of Indian agriculture	2011–2014	100 climatically vulnerable villages	Incorporated NICRA experience (while undergoing expansion via replication) within the National Mission on Sustainable Agriculture
12	NeGPA	Increase farm productivity	2011–2015	7 states	Scaled-up to all states and bundled with a new e-kranti or Digital India initiative aiming at digitizing relevant services for the citizens for better governance and public service delivery
13	National Agriculture Technology Project (NATP)	Research and extension reforms	1998–2005	28 districts in 7 states	Extended NATP model to all districts in the country and a new Centrally Sponsored Scheme on Support to State Extension Programmes for Extension Reforms was initiated
14	Accelerated Fodder Development Programme	Facilitate additional production of fodder in the country	2011–2015	12 states	Subsumed under National Food Security Mission

172

of the pilot (pilot outcome). Set-theoretic methods can be deployed if the phenomenon or concept of interest is best studied in terms of set relations (Schneider and Wagemann, 2012). Scaling up as a phenomenon is understood to be a result of such a combination or set of factors (Gillespie, 2004; Hartmann and Lin, 2007; Simmons, Fajans and Ghiron, 2007; Vreugdenhil et al., 2012;).

One of the most formalized techniques for set-relational research (Schneider and Rohfling, 2013) is qualitative comparative analysis (QCA), as initially developed by Prof. Charles C. Ragin and colleagues at the University of Arizona. The objective of QCA is to enable causal interpretation in addition to detailed qualitative information that is obtained from case studies in order to understand the different combination of plausible factors that could lead to a specific outcome (Ragin, 2006; 2007). This analysis is based on fuzzy-set QCA (fsQCA) wherein variables are scored between 0 and 1 (Ragin, 2006). QCA helps bring a level of abstraction to study concepts and phenomenon that are otherwise dependent on highly context-driven single case analysis studies (Legewie, 2013). fsQCA is particularly helpful in instances where there is a proposition/hypothesis regarding the underlying causal factors affecting the outcome being studied (scaling up in this case) (Ragin, 2007; Kent, 2008).

To explore whether the implementation of a new pilot led to a policy mix with coherent goals and consistent means, the goals and instruments of each pilot and changes brought over during the pilot phase to an incumbent regime were traced through the use of semi-structured interviews with key informants and policy document analysis (similar to Kern and Howlett, 2009).

While fuzzy-set scoring for a condition or outcome can move from a scale of 3 to more than 6 divisions between the values of 0 and 1 (Ragin 2006, 2007), a 4-point scale (i.e. 0, 0.33, 0.67 and 1) is adopted for this study. This scoring logic is based on theory (Rondinelli, 1993; Vreugdenhil and Slinger, 2012; Vreugdenhil et al. 2012) as well as substantiated with empirical evidence from the field interviews and document analysis. Goertz's (2006) concept is followed to study the outcome in this study, i.e. scaling up as a continuum moving from termination of pilot in the same form (0) to full institutionalization (1). On one extreme is no policy integration (score 0) and on the other extreme is full institutionalization in the form of reforms (score 1), i.e. national new policies initiated based on the pilot. The fuzzy area between the two extremes of the continuum is then theorized. Following Vreugdenhil and Slinger (2012), Vreugdenhil et al. (2012) and Rondinelli (1993), the fuzzy area in between 0 and 1 is scored as 0.33 = expansion into multiple pilots via demonstration in different contexts but at the same scale (Rondinelli, 1993) and 0.67 = expansion of the pilot with replication and scale changes, more administrative layers added and bundling of the pilot in ongoing schemes as a different scale.

A scoring scheme is developed to classify and study the changes in elements of an existing policy mix, based on the Cashore and Howlett (2007) and Hall (1993) schemes cited previously. The extremes for the three elements at the policy goal level move from no changes in policy goals to complete change from the respective status quo policy regime. Similarly, the extremes for the three elements at the policy means level move from no change in the means to complete change from status quo policy regime (Table 9.2 and Table 9.3).

Table 9.4 shows the data matrix for the fourteen case studies, outcome and the four causal conditions.[2]

Note: ECIS: Experimental Crop Insurance Scheme; FIIS: Farm Income Insurance Scheme; WBCIS: Weather Based Crop Insurance Scheme; MNAIS: Modified National Agriculture Insurance Scheme; NWDPRA: National Watershed Development Programme for Rainfed Areas; IGWDP: Indira Gandhi Watershed Development Programme; NPOF: National Project on Organic Farming; RADP: Rainfed Area Development Programme; NAIP: National Agriculture Innovation Project; NICRA: National Initiative for Climate Resilient Agriculture; NeGPA: National e-Governance Plan in Agriculture; NATP: National Agriculture Technology Project; AFDP- Accelerated Fodder Development Programme

Of the fourteen cases, only one, i.e. National Agriculture Technology Project (NATP), was fully institutionalized. Ten gave rise to new management programmes or were bundled with existing policies and programmes. Two pilots remained in the replication phase (one of which, Farm Income Insurance Scheme, terminated soon after). Only one pilot (Experimental Crop Insurance Scheme) was a failure in terms of not moving beyond the pilot phase, and was terminated within one season.

Results of the QCA indicate that only pilots involving marginal changes to the status quo or current policy mix were successfully scaled up in the agriculture sector in India, with bulk being merged and one being institutionalized.[3] That is, the majority of the pilots scaled up can be subsumed under or bundled with ongoing government schemes and programmes, becoming part of the existing policy mix.

Three combinations are found to be sufficient to explain scaling up[4] (details in Appendix). The principal combination ~obj*~calibration indicates a situation where there is no change in objectives and no change in instrument calibration. This reflects that for both the political level (goals) and the operationalization level

---

[2] Denoting changes in objectives, settings, instrument type and instrument calibration respectively.

[3] Although incremental changes over time may lead to major policy changes, the study of pilots cannot capture these gradual changes owing to their limited time period of operation.

[4] ~ indicates negation or absence of a particular causal condition. Thus ~setting suggests no change brought by the pilot at the settings level of an existing policy.

Table 9.2. *Scoring logic for the four conditions*[a]

Score	Logic	Change in goals		Change in means	
		Change in objectives	Changes in settings	Changes in instrument logic and type	Changes in instrument calibration
0	No change	The pilot does not suggest any change to the current policy objectives.	The pilot does not suggest any change in on-ground requirements to meet the objectives.	The pilot works with the same instrument logic and type of instrument(s) as in the previous policy/ programme.	The instruments are utilized in the same manner as the earlier policy/programme.
0.33	Classic incremental	The pilot brings an incremental addition to the same objectives (same direction, no conflict).	The pilot brings an incremental addition to the same on-ground requirements of the policy/ programme (same direction, no conflict).	The pilot adds new instruments as an incremental addition to the same type of instruments (same direction, no conflict, more instruments of the same type).	The pilot brings incremental changes to the same way in which instruments are utilized (same direction, no conflict).
0.67	Contested incremental	Completely new objectives are added to old ones, sometimes leading to conflict.	Completely new on-ground policy requirements are introduced, moving the programme towards a new equilibrium, some of which might be in conflict with existing ones.	The pilot maintains old instrument types while adding new, but some of these might conflict with existing ones.	The pilot maintains old instrument settings while new are added; some of these settings might conflict with existing ones.
1	Paradigmatic change	Completely new objectives are introduced, dismantling the previous ones.	The pilot puts forth completely new settings to operationalize objectives, dismantling old ones.	The pilot puts forth completely new instruments while removing old ones.	The pilot puts forth completely new instrument settings while dismantling old ones.

[a] Based on the data collection, two changes were made to the model. Firstly, goals were not found to change. That is, none of the pilots sought to change the overall abstract goals of an incumbent policy or programme; hence, it was 0 throughout for all the cases. Thus the Change in Goals variable was dropped from the model. Secondly, the variable/condition Instrument Logic guided the type of instrument that was to be used. Thus, the scores on both completely matched across all the cases; hence Instrument Logic and Type were combined to represent one variable called Change in Instrument Type. Thus, the model proceeded for analysis with a total of four causal conditions (changes in policy objectives, settings, instrument type and instrument calibration), from an initial set of six conditions.

Table 9.3. *Scoring scheme for outcome (scaling up)*

Score	Outcome	Criteria for scoring	Implications
0	No scaling up	Termination of pilot without replication, expansion or institutionalization	Pilot has not been scaled up; it finishes its term in the same form or has an early termination.
0.33	Limited scaling up (replication)	Generation of multiple comparable pilots in other locations and over time	The pilot model in the same form is being replicated in multiple similar contexts (similar in scale, complexity and policy issue addressed).
0.67	Substantial scaling up (expansion)	Expansion of the pilot itself, lessons drawn from the pilot to initiate new management project(s) (Vreugdenhil et al., 2012) or bundling of the pilot with ongoing programmes/ policies	Scope of the pilot expands including its structure and functions, implementation context, stakeholder groups and administrative layers through direct expansion, bundling or development of new management projects.
1	Full scaling up (institutionalisation)	Full institutionalization or development of new national/regional policies based on the pilot	The knowledge generated through the pilot becomes part of the standard operating procedure in specific government policies.

Table 9.4. *Fuzzy scores assigned for conditions and outcome*

Pilot	Outcome	Objectives	Setting	Instrument type	Calibration
ECIS	0	0.33	0.67	0	1
FIIS	0.33	1	1	0	1
WBCIS	0.67	0.67	1	0	1
MNAIS	0.67	0.33	0.67	0	0.33
NWDPRA	0.33	0.33	0.33	0.33	0.33
IGWDP	0.67	0.33	0.33	0.67	0.33
SUJALA	0.67	0.33	0.33	0.33	0.33
NPOF	0.67	0.33	0.67	0.67	0.33
RADP	0.67	0.33	0.33	0.33	0.67
NAIP	0.67	0.33	0.33	0.67	0.67
NICRA	0.67	0.33	0.67	0.67	0.33
NeGPA	0.67	0.33	0.33	0.33	0.33
NATP	1	0.67	0.33	0.33	0.67
AFDP	0.67	0.33	0.33	0.33	0.33

(state and sub-state level), as long as there is no change in objectives or calibration of policy elements involved, pilots are likely to scale up. The second combination [~obj*~setting] indicates a situation where there is no change in objectives and no change in settings. It again hints towards the political nature of scaling up. So, as long as the objectives and settings remain the same, scaling up will happen irrespective of small or major changes at the operationalisation level. The third sufficient combination [~setting*~itype*calibration] is a bit more complex and indicates a situation of no change in settings and instrument type but substantial change in instrument calibration.

9.3 Discussion and Conclusion

Policy piloting has been suggested as a useful instrument for governments to test new policies by initiating them on a small scale first, thereby, testing their acceptability to the beneficiaries and stakeholders at large and, subsequently, followed by incremental changes or large-scale reforms to existing plans and schemes. The goal of such policy pilots is to guide future policy development. Piloting holds particular importance for complex and wicked problems such as climate change adaptation where policy responses are being planned under a high level of uncertainty. The questions of whether and how such pilots eventually scale up and translate into policy have been of interest to policy scholars and practitioners alike. Pilots successful at a small scale can be continued or expanded in various forms; however, there is little empirical analysis on the factors leading to their diffusion and the process of diffusion as well as their actual contribution to altering the policy mix. For example, apart from the technical strengths of pilots, sometimes these may be scaled up for political reasons including their degree of compliance with status quo or the current policy mix.

The case analysis reveals that even though the federal Ministry of Agriculture, Government of India, may not consider piloting as a regular feature of policy formulation and change, piloting is leading to innovation and learning to improve current policy mixes. The study and its case analysis have several implications for the design of policy pilots, their scaling up and the emergence of new policy mixes under conditions of risk and uncertainty. Firstly, multiple pathways to scaling up of policy pilots were observed. Secondly, the combination of changes at the policy ends and means level associated with successful pilots were found to be rather conservative, that is, characterized by incremental adjustments to the current policy mix.

Thirdly, the outcome of majority of the pilots was found to be a merger with other ongoing policies and programmes to increase their current scale and scope. The tendency to adhere to slow and gradual adjustments in policy components also indicates towards the need to be judicious with current resource investment profiles in the agriculture sector. Thus, despite the theoretical acknowledgment as an approach that can enable risk-taking and experiment with policy alternatives under uncertainty, in practice the operational contribution of the studied policy pilots was found to be limited to acting as avenues for periodically updating existing policy mixes through marginal changes to their current scope.

The results of the study in the Indian agriculture context highlight a disconnect between the theoretical importance bestowed on policy pilots and experiments as a useful approach towards risk management, investments in policy alternatives and innovations and scaling up and policy integration of these pilots in practice (see also van Buuren et al., Chapter 8). Especially in the climate change adaptation context, it can translate into a preference for continuing status quo activities under high levels of uncertainty related to climate change impacts.

References

Beland, D. (2007). Ideas and institutional change in social security: Conversion, layering, and policy drift. *Social Science Quarterly*, 88 (l), 20–38.

Cabinet Office (2003). Trying it out: The role of 'pilots' in policy-making. In *Report of a Review of Government Pilots*. London: Strategy Unit, Government of the United Kingdom.

Cashore, B., and Howlett, M. (2007). Punctuating which equilibrium? Understanding thermostatic policy dynamics in Pacific Northwest Forestry. *American Journal of Political Science*, 51, 532–551.

Christiansen, L., Olhoff, A., and Trærup, S. (eds.) (2011). *Technologies for adaptation: Perspectives and practical experiences*. Roskilde: UNEP (United Nations Environment Programme) Risø Centre.

Deyle, R. E. (1994). Conflict, uncertainty, and the role of planning and analysis in public policy innovation. *Policy Studies Journal*, 22(3), 457.

Gillespie, S. (2004). *Scaling Up Community-Driven Development: A Synthesis of Experience*. International Food Policy Research Institute. Washington D.C.

Goertz, G. (2006). *Social Science Concepts: A User's Guide*. Princeton, NJ: Princeton University Press.

GoI (Government of India) (2007). *Report of the Working Group on Risk Management in Agriculture for XI Five Year Plan (2007– 2012)*. New Delhi: Planning Commission, Government of India.

GoI (2011). *Report of the XII Plan Working Group on Natural Resource Management and Rainfed Farming*. New Delhi: Planning Commission, Government of India.

Hall, P. A. (1993). Policy paradigms, social learning and the state: The case of economic policymaking in Britain. *Comparative Politics*, 25(3), 275–296.

Hartmann, A., and Linn, J. F. (2007). Scaling Up: A Path to Effective Development. In *2020 Focus Brief on the World's Poor and Hungry People*. Washington, DC: IFPRI (International Food Policy Research Institute)

Hoffmann, M. J. (2011). *Climate Governance at the Crossroads: Experimenting with a Global Response after Kyoto*. Oxford: Oxford University Press.

Howlett, M., and Cashore, B. (2009). The dependent variable problem in the study of policy change: Understanding policy change as a methodological problem. *Journal of Comparative Policy Analysis: Research and Practice*, 11(1), 33–46. http://doi.org/ 10.1080/13876980802648144

Howlett, M., & Rayner, J. (2007). Design principles for policy mixes: Cohesion and coherence in 'new governance arrangements'. *Policy and Society*, 26(4), 1–18.

Howlett, M., and Rayner, J. (2013). Patching vs packaging in policy formulation: Assessing policy portfolio design. *Politics and Governance*, 1(2), 170–182.

Jordan, A., and Huitema, D. (2014). Innovations in climate policy: Conclusions and new directions. *Environmental Politics*. doi: 10.1080/09644016.2014.924209. 906-925.

Kent, R. (2008). *Using fsQCA: A Brief Guide and Workshop for Fuzzy-Set Qualitative Comparative Analysis*. United Kingdom: University of Stirling.

Kern, F., and Howlett, M. (2009). Implementing transition management as policy reforms: A case study of the Dutch energy sector. *Policy Sciences*, 42, 391–408.

Kwadijk, J. C. J., Haasnoot, M., Mulder, J. P. M., Hoogvliet, M. M. C., Jeuken, A. B. M., et al. (2010). Using adaptation tipping points to prepare for climate change and sea level rise: A case study in the Netherlands. *Wiley Interdisciplinary Reviews Climate Change*, 1(5), 725–740.

Legewie, N. (2013). An introduction to applied data analysis with qualitative comparative analysis (QCA). *Qualitative Social Research*, 14(3). http://nbn-resolving.de/urn:nbn: de:0114-fqs1303154. Accessed on 23 February, 2016.

Majone, G. (1991). Cross-national sources of regulatory policymaking in Europe and the United States. *Journal of Public Policy*, 11, 79-106.

O'Brien, K., Pelling, M., Patwardhan, A., Hallegatte, S., Maskrey, A., Oki, T., Oswald-Spring, U., Wilbanks, T., and Yanda, P. Z. (2012). Toward a sustainable and resilient future. In Field, C. B., Barros, V., Stocker, T. F., Qin, D., Dokken, D. J., Ebi, K. L., Mastrandrea, M. D., Mach, K. J., Plattner, G.-K., Allen, S. K., Tignor, M., and Midgley, P. M. (eds.), *Managing the Risks of Extreme Events and Disasters to Advance Climate Change Adaptation*. A Special Report of Working Groups I and II of the IPCC. Cambridge: Cambridge University Press, 437–486.

Parry, M. L., Canziani, O. F., Palutikof, J. P., van der Linden, P. J., and Hanson, C. E. (eds.) (2007). *Contribution of Working Group II to the Fourth Assessment Report of the Intergovernmental Panel on Climate Change, 2007*. Cambridge: Cambridge University Press.

Ragin, C. (2006). Set relations in social research: Evaluating their consistency and coverage. *Political Analysis* 14, 291–310.

Ragin, C. (2007). Qualitative comparative analysis using fuzzy sets (fsQCA). In Rihoux, B., and Ragin, C. (eds.), *Configurational Comparative Analysis*. Berkeley, CA: Sage Publications, 87–121.

Rip, A., and Kemp, R. (1998). Technological change. In Rayner, S., and Malone, L. (eds.), *Human Choice and Climate Change*. Vol. 2. *Resources and Technology*. Washington, DC: Batelle Press, 327–399.

Rondinelli, D. (1993). *Development Projects as Policy Experiments: An Adaptive Approach to Development Administration*. 2nd edn. New York: Routledge.

Schneider, C. Q., and Wagemann, C. (2012). *Set-Theoretic Methods for the Social Sciences*. Cambridge: Cambridge University Press.

Schneider, C. Q., and Rohlfing, I. (2013). Combining QCA and process tracing in set-theoretic multi-method research. *Sociological Methods & Research*, 42(4), 559–597. http://doi.org/10.1177/0049124113481341

Simmons, R., Fajans, P., and Ghiron, L. (eds.) (2007). *Scaling Up Health Service Delivery: From Pilot Innovations to Policies and Programmes*. New York: World Health Organization.

Swanson, D., and Bhadwal, S. (eds.) (2009). *Creating Adaptive Policies: A Guide for Policymaking in an Uncertain World*. New Delhi: Sage Publications, IDRC.

Thelen, K. (2003). How institutions evolve: Insights from comparative historical analysis. In Mahoney, J., and Rueschemeyer, D. (eds.), *Comparative Historical Analysis in the Social Sciences*. Cambridge, MA: Cambridge University Press, 208–240.

Vreugdenhil, H., Frantzeskaki, N., Taljaard, S., Ker Rault, P., and Slinger, J. (6–8 April 2009). The next step in policy transitions: Diffusion of pilot projects. 13th Annual Conference of the International Research Society for Public Management, Copenhagen Business School, Denmark.

Vreugdenhil, H., Taljaard, S., and Slinger, J. H. (2012). Pilot projects and their diffusion: A case study of integrated coastal management in South Africa. *International Journal of Sustainable Development*, 15(1–2), 148–172.

Weiss, C. (1975). Evaluation research in the political context. In Struening, E. L., and Guttentag, M. (eds.), *Handbook of Evaluation Research*. Thousand Oaks, CA: Sage Publications, 13–26.

Wilder, M., and Howlett, M. (2014). The politics of policy anomalies: Bricolage and the hermeneutics of paradigms. *Critical Policy Studies*, 8(2), 183–202. http://doi.org/10.1080/19460171.2014.901175

Appendix

Table A.1. *Analysis of sufficiency conditions*

	~obj*~calibration	~obj*~setting	~setting*~itype*calibration
Raw coverage	0.719	0.678	0.516
Unique coverage	0.122	0.04	0.04
Consistency	0.946	0.894	0.929
Solution coverage	0.84		
Solution consistency	0.913		

The results of a QCA are interpreted through the following key concepts (Ragin, 2006, Ragin, 2007):

Necessary condition: A causal condition (X) is considered necessary for outcome (Y), if Y cannot occur without X, i.e. Y (outcome) is a subset of X (cause).

Sufficient condition: If a causal condition (X) is sufficient for outcome (Y), then, if Y is present, X must be present too. However, there may be other factors leading to Y, not only X. Here, X (cause) is a subset of Y (outcome).

Consistency assesses the degree to which the cases sharing a given combination agree in displaying the outcome.

Coverage assesses the degree to which a cause or causal combination accounts for instances of an outcome. When there are several paths to the same outcome, the coverage of any given causal combination may be small, thus coverage gauges empirical relevance or importance.

Solution coverage measures the proportion of memberships in the outcome that is explained by the complete solution (A*B + A*C, etc.).

Raw coverage measures the proportion of memberships in the outcome explained by each term of the solution (A*B).

Unique coverage measures the proportion of memberships in the outcome explained solely by an individual solution term (memberships not covered by other solution terms).

10

Evaluating Climate Governance Experiments

Participants' Perspectives on Low-Carbon Experiments in Finland

EVA HEISKANEN AND KAISA MATSCHOSS

10.1 Introduction

Experiments have gained significant traction in climate governance. Yet the arguments made for experimentation are quite diverse. Experimentation may be a sound philosophical stance under conditions of fundamental uncertainty (Overdevest, Bleicher and Gross, 2010), or a concrete tool for transition management, with experiments materialising and systematically revising a selected transition pathway (Kemp, Schot and Hoogman, 1998). Experiments may be seen as a way to delegate power as a part of open policymaking (Sabel and Zeitlin, 2008) or as a way to make policy more effective by piloting interventions and using experimental evidence to select the most appropriate interventions (Pearce and Raman, 2014). In public discourse, these different meanings of experimentation are often merged into a grand narrative of entrepreneurial and adaptive policymaking (e.g. Mulgan, 2014).

Local sustainability experimentation can make an important contribution to experimentalist climate governance. Many such experiments arise from local interests to find ways to solve climate challenges outside existing political hierarchies (Hoffmann, 2011) by developing new urban infrastructure, trying out new energy solutions in the built environment, creating new transport systems, or experimenting in climate change adaptation (Bulkeley and Castán Broto, 2013). Such initiatives have the capacity to provide valuable lessons for national or pan-national governance of sustainability transitions (Schot and Geels, 2008). On the other hand, living labs and other urban experiments explicitly employ the local context as a site for experimentation with new technology in the hands of real users (Evans, 2011; Evans and Karvonen, 2014; Voytenko et al., 2016). When public funding is devoted to such experiments, they can serve the dual purposes of promoting local aims while contributing to wider technological or governance developments (e.g. Hodson and Marvin, 2009).

Learning is central for the expansion and embedding of experiments. In local sustainability experimentation, however, learning is complex for several reasons. First, local sustainability experiments combine several technologies, infrastructures and social systems – old and new (Neass and Vogel, 2012). Second, they engage diverse actors and diverse interests – only some related to learning (Quitzau, Hoffmann and Elle, 2012; Bulkeley and Castán Broto, 2013; Evans and Karvonen, 2014). Third, learning in the sense of embedding can occur locally (improved local capacity), it can transfer to other sites (Raven et al., 2008), or it can be aggregated and analysed across sites. This third type of learning would be important in order for experiments to contribute to 'social learning on how to advance climate change mitigation' rather than mere advocacy of a particular solution (Kivimaa, Hilden, Huitema et al., 2017).

When experiments are considered as a form of climate governance, evaluation becomes important for learning and legitimacy (Jordan and Huitema, 2014; Kivimaa et al., 2017). From the (sustainable) technology proponents' perspective, sustainable niches can scale-up via various processes (Schot and Geels, 2008), but from a state policy perspective, such advocacy might not be a legitimate aim for experimentalist governance (De Schutter, 2010). Moreover, if governance is to be transformed and enhanced via experimentation, there might also be a need to aggregate negative – alongside positive – lessons (Kivimaa et al., 2017).

This chapter draws on two perspectives on learning from experiments: the learning perspective of strategic niche management (SNM) and some of its qualifications and programme evaluation perspectives which are used to learn from policy interventions. We illustrate this framework through two local sustainability experiments that draw their momentum from local interests but are also explicitly part of national-scale attempts to learn about viable technological configurations. Our cases are two emblematic local experiments from Finland, a country that has put experimentation explicitly on the government's agenda: Kalasatama in Helsinki, a national smart grid and built environment pilot project; and Carbon-Neutral Municipalities, which engages municipalities in experimentation for reducing greenhouse gas emissions by 80 per cent by 2030. Our empirical data focuses on how participants interpret local climate policy experiments: what (in their view) the experiments are testing, and what can be learned from the outcomes.

Through these illustrations, we reflect on the problem of evaluating local sustainability experiments and the lessons that can be drawn from them for climate governance. Our central thesis draws on constructivist evaluation: the 'solutions' being tested and their 'context' are inextricable (Dahler-Larsen and Schwandt, 2012), and therefore learning is by necessity situated (Holm, Stauning and Søndergård, 2011; Carvalho and Lazzerini, Chapter 3). Hence, evaluation of mere 'solutions', especially in a governance context, is not straightforward. In the concluding section, we suggest some ways forward in order to facilitate the embedding of the lessons learned from experiments beyond their local origins.

10.2 Climate Governance: Perspectives on Learning from Local Experiments

To understand potential processes by which governance experiments might grow out of their initial context (Turnheim, Kivimaa and Berkhout, Chapter 1) by analysing the lessons learned, we next engage with two perspectives from quite different literatures. We first discuss how SNM conceptualises learning from local experiments. We then turn to approaches to learning from policy interventions in theory-based and constructivist evaluation.

10.2.1 Strategic Niche Management

The SNM perspective (Schot and Geels, 2008) builds on a quasi-evolutionary model of how experiments contribute to wider technological learning. It conceives of local experiments as sites for technological development via variation, selection and retention within protected spaces (Geels and Raven, 2006; Raven et al., 2008). Learning in strategic niches is about developing new technical solutions and capabilities; developing new supply chains; adapting the new solutions to regulations, infrastructures, markets and user needs; and developing new cultural meanings surrounding the solutions, as well as learning about societal and environmental effects. It can be single-loop learning (acquisition of facts and skills), double-loop learning (changes in expectations and assumptions) or second-order learning (learning to learn) (cf. Argyris, 2003; Hoogma, Weber and Elzen, 2005; Rohracher, 2008).

Bluntly put, from this perspective, experiments are successful if they are selected and retained for further development – that is, if the technologies and related social arrangements are continued and expanded beyond their original (protected) site. Such processes of selection and retention involve learning not only by the developers of the technology but also by the users and other stakeholders implicated by its use. Selection entails mutual alignment of expectations: the implementation of the project also changes the context (Raven et al., 2008). For the wider community of sustainable technology advocates, each project yields new results, which can be aggregated into generic lessons for the emerging trajectory of the sustainable technology (Geels and Deuten, 2006; Raven et al., 2008). Failures contribute to the overall processes of niche and regime shaping, as well as altering environmental selection rules. Therefore, not only individual 'successful' initiatives have positive value from a systemic perspective.

10.2.2 Local Politics of Experimentation

Local experiments might not serve merely, or even mainly, as open-ended processes of learning. Bulkeley and Castán Broto (2013) extend the concept of experimentation beyond that of learning via trial and error: experimentation is used to advance divergent values and claims and to establish or contest various forms of authority and control. For example, Hodson and Marvin (2007) have highlighted the contested nature of local experiments in which local residents may resist serving as a test bed for external corporate interests, whereas Evans and Karvonen (2014) show how knowledge-producing local experiments can serve to reinforce the dominance of those in power.

Although experiments may be about testing new technologies in field conditions (i.e. what works when and where), they may also be about creating new forms of governance. They may also be fundamentally performative about materialising and visualising particular visions of the future (Blok, 2013). From this perspective, the evaluation of experiments unavoidably involves several actors with different expectations and interpretations of what outcomes are successful or desirable (Raven et al., 2016).

10.2.3 Perspectives on Learning from Realist and Constructivist Evaluation

If local experiments are to contribute to a form of experimentalist governance in climate policy, they should support learning from implementation and periodic re-evaluation of goals and practices (Sabel and Zeitlin, 2008; De Búrca, Keohane and Sabel, 2014). Two streams of research from the evaluation literature seem relevant for the form of networked governance that characterises local experiments: participatory evaluation, especially 'theories of change', and realistic evaluation, which both aim to identify how and why policy interventions work within their context of application (Blamey and McKenzie, 2007).

The 'theories of change' literature (Blamey and McKenzie, 2007) starts from the notion that participants involved in the implementation of an intervention hold different views of what a policy intervention is doing and what might be the underlying 'mechanisms' that make it work. Part of the evaluator's work is to facilitate the articulation of these theories and to highlight potential discrepancies between them (Blamey and McKenzie, 2007). Hertting and Vedung (2012) argue that this is particularly the case in the present-day network governance mode, i.e. when governance is not top-down but employs horizontal coordination, crosses administrative boundaries and public vs. private spheres and uses informal control. Under such conditions, Hertting and Vedung (2012) argue that evaluation should capture actors' individual goals in their diversity and potential contradictions, as well as emergent common goals of the governance network.

Capturing and problematising participants' theories of change might not always be easy, however. Sometimes, stakeholders might not want to recognise the causes of success or failure (Vedung, 2006). Friedman (2001) has discussed 'designed blindness' in evaluation: stakeholder groups may actively overlook contradictions between the viewpoints of different groups and inconsistencies within their own views to avoid conflict and to distance themselves from failure.

Realistic evaluation has a strong interest in drawing lessons across individual evaluations to learn how the working of the 'mechanism' of an intervention is dependent on particular contexts (Pawson and Tilley, 1997). Yet it differs from positivist randomised control trials by acknowledging that the context is central to the outcome of the intervention. It aims to uncover the relationship between the 'mechanism' that the programme aims to test and a certain 'outcome' in different 'contexts', recognising that 'mechanisms' work differently in different contexts and that the context can be part of the reason why a programme works (Pawson and Tilley, 1997). Since realistic evaluation acknowledges that contexts influence outcomes, this perspective is compatible with current ideas of 'governance by experimentation', where the deployment of ideas in real-life contexts is their ultimate 'test'.

However, realistic evaluation struggles with the fact that stakeholders' views on an intervention may also influence the performance of the intervention. For example, Dahler-Larsen (2001) has discussed how programmes can create enthusiasm and support in their context and be successful even if the underlying mechanism tested does not work ('magic programmes') or, conversely, change the context in ways that undermine the working of sound mechanisms ('tragic programmes'). This is because programmes can modify their context, for example, by undermining the confidence of participants, or by creating a 'positive spin' surrounding the programme. This is because the 'mechanism' that an intervention aims to test cannot be separated from its 'context' (Dahler-Larsen, 2001).

10.2.4 Summary and Framework for Analysis

The literatures previously presented offer ingredients for analysing the embedding of local sustainability experiments within and beyond their context (Table 10.1). Though stemming from diverse sources, together they help focus on a combination of 'solution' (niche technology in SNM, mechanism in realistic evaluation) and 'context' that makes an experiment 'work'. Moreover, they bring in viewpoints of diverse participants (political perspectives) and acknowledge that experiments can 'work' in different ways, irrespective of the viability of the solutions tested (constructivist evaluation).

Table 10.1. *Ingredients for analysing the embedding of local sustainability experiments*

Strategic niche management	Realistic evaluation and theories of change
• Learning via variation, selection and retention through experimentation in protected spaces • Aggregation of niche lessons in emerging global (single-)niche community	• Experiments are about testing diverse combinations of mechanisms and contexts (which stakeholders understand differently) • Learning occurs by identifying context-mechanism-combinations that leads to desired outcomes
Political perspective on local experimentation	**Constructivist evaluation**
• Local experiments are contested over the direction of local development. • Local experiments involve contests over dominance locally and beyond.	• Experiments are performative: they change the context. • Experiments influence their conditions of success (also irrespective of the mechanisms) by creating or undermining commitment and confidence.

Our aim in this illustration is not to evaluate the selected case experiments, or to test theoretical propositions, but to explore the possibilities and limitations of such an evaluation. Following the 'theories of change' perspective, we are interested in discovering what 'solutions' the participants think the experiment is testing, and how and why they feel the experiment has been successful. In this, we draw on the view advocated by Hertting and Vedung (2012) that experiments represent networked modes of governance in which different parties engage for different reasons. We define 'participants' as people directly impacted by the experiment – the people 'inside' – and stakeholders as actors who have an interest and are involved in the experiment (such as e.g. city authorities) but who are 'outsiders' to the experiment (although, in practice, the roles are more blurred). Together they are called 'actors'. We also acknowledge that actors' interests and expectations can change with time (as is expected in the SNM approach) and that actors' interests can also be emergent.

We investigate learning from several perspectives, building on what the actors themselves see as learning, and try to trace possible ways in which lessons learned are mobilised beyond the local context. We also aim to uncover why participants are likely to make particular kinds of judgement, which we do by comparing diverse participants' viewpoints and by analysing how their judgements relate to their experience of the experiment from their particular position. Through this framework, we attempt to explore the following questions:

• What do participants think the experiment is 'testing'?
• What kinds of learning occur in local sustainability experiments?

• How is learning aggregated across sites?
• How might local sustainability experiments be evaluated, and what problems are likely to be encountered in such evaluations?

In the following, we apply this framework of analysis to two cases that explicitly present themselves as local sustainability experiments in Finland, a country that has recently made a commitment to develop an 'experimental culture' of governance (Government Programme, 2015: 28).

10.3 Two Emblematic Experiments in a Rural and Urban Context

We have chosen for this illustration two cases that represent a prevalent form of local sustainability and low-carbon experimentation in Europe, insofar as local governments are key initiators and their main focus is in developing low-carbon energy solutions in the built environment (Bulkeley and Castán Broto, 2013). However, the contexts of these experiments are almost polar opposites: one is about developing a new and innovative urban structure in the middle of the capital city, whereas the other is about renewing existing structures toward climate neutrality in small rural municipalities. This dichotomy aims to capture participants' potentially diverse views on the role of 'experimentalism' in everyday life, while any common features found might be expected to be found across other cases falling between these poles.

10.3.1 Case 1: Smart City Development in Kalasatama, Helsinki

The new Kalasatama area of Helsinki is an experimental innovation platform to co-create smart urban infrastructure and services in close co-operation with residents, city officials and other stakeholders such as industry, SMEs and researchers through piloting of novel low-carbon energy solutions. The vision of Smart Kalasatama is to become so resource-wise that residents will gain an extra hour of own time every day. Smart Kalasatama has a wide scope, covering a variety of different industries and solutions ranging from smart grids to eHealth and smart retail. It is an urban living lab, forming an open innovation platform that offers a place to co-create new urban services with users in a real environment.

Smart Kalasatama-project has initiated a Programme for Agile Piloting, which is based on an idea to launch a call for new pilot ideas providing new innovative services for people living in the Kalasatama area, purchase the most interesting ones (1,000–8,000 euros) and launch them quickly. The aim of the programme is to accelerate ideas to service innovations and reach users in a real life setting as well as spur experiments produced by several actors in a protected niche. The first pilots started during the spring 2016. The Developers' club, on the other hand, aims at networking different actors (such as bigger and smaller firms, city officials and residents) active in Kalasatama to enable a better information flow about future events, changes and plans, which makes finding partners for cooperation and planning joint projects easier. They adjourn four times a year. The Developers' club presents a novel way of cooperation at a city district level, which offers an easy way of bringing new experiments into the area.

During 2015–2017 Smart Kalasatama is part of a spearhead project for a strategy for sustainable urban development by the six largest cities in Finland. Smart Kalasatama is funded by the European Regional Development Fund, the City of Helsinki and the Ministry of Employment and the Economy. The project is coordinated by Forum Virium Helsinki, which is an innovation intermediary owned by the City of Helsinki. By the early 2030s, the district is expected to offer a home for approximately 20,000 residents and create jobs for 8,000 people. Currently, 2,000 people are living in the area.

10.3.2 Case 2: Carbon-Neutral Municipalities (CANEMU) in Mynämäki

CANEMU was launched in the autumn of 2008 with the aim to engage munici-palities outside the metropolitan area as 'change laboratories' for new solutions to climate change. The initiative arose from co-operation between a business leader's social responsibility initiative and the Finnish Environment Institute. They origin-ally selected five small municipalities to partner with (by 2016, there are more than thirty participating municipalities). The municipalities joining the project pledged to decrease greenhouse gas emissions from 2007 levels by 80 per cent by 2030. For this, they gained technical support from the Finnish Environment Institute and received a little project funding from Tekes, the Finnish Funding Agency for Technology and Innovation, and the participating companies. CANEMU has stressed a bottom-up process of change, where solutions are sought for and tested together with local citizens, businesses and municipal administrations (Mickwitz et al., 2011).

We focus on one of the five pioneering municipalities, Mynämäki, with about 8,000 inhabitants dispersed over a wide area. The municipality consists of a central village and other more built-up areas, as well as open countryside with farms and forests. Relative to the rest of the country, the community hosts a large number of farms (>300) and small businesses (>450). In Mynämäki, the CANEMU experi-ment has focused on promoting energy efficiency solutions (LEDs, renovations, energy monitoring and energy town plan) in public and private buildings, as well as shifting from oil and electricity to renewable energy: ground source heat pumps, bioenergy, solar collectors and solar power. Innovative ideas have included, e.g. village joint heating systems, shared workspace and the production of energy from waste heat and agricultural residues, as well as joint purchasing of solar panels (Heiskanen et al., 2015).

10.3.3 Data Collection and Analysis

Participants' and funding bodies' viewpoints were collected with open-ended interviews, informed by the conceptual framework. Some of the interviews had a somewhat different focus originally but were sufficiently open-ended to be used for investigating the problems of interest in this article. Table 10.2 outlines the types of interviewees represented in the data. Additionally, documents, websites and newspaper articles were analysed. To supplement the potential aggregators' perspective, we draw on notes taken at a workshop organised for national-level authorities and experts, with a total of twelve participants. In our analysis, we focus more on similarities than differences, using these two extreme cases to identify themes that are likely relevant across experiments.

10.4 Empirical Illustration: What Is Tested, What Is Learned and What Influences Judgements

10.4.1 Diverse Understandings of What Is Being Tested

Local sustainability experiments are diverse and complex. Governance experiments organised under the auspices of local government tend to experiment with combinations of mechanisms, such as hands-on advice or feedback, new technologies including solar panels or energy monitoring and control devices and new types of urban infrastructures like shared transports. These are combined with contextual elements such as buildings, streets and roads and business models. They involve actors, such as residents, developers, businesses and local government administrations. Given this diversity of 'experiments within the experiment', it is natural that different actors have *diverse or even divergent understandings* of what is

Table 10.2. *Key informants for the case studies*

Smart Kalasatama	CANEMU in Mynämäki
Interviews with experiment organisers: 3	Interviews with initiators/organisers: 4
Interviews with local politicians and officials: 5	Interviews with local politicians and officials: 6
Interviews with residents: 15	Interviews with local residents: 20
Interviews with funding bodies/potential aggregators or lessons: 2	Interviews with funding bodies/potential aggregators of lessons: 4

Table 10.3. *Diverse perspectives on what 'solutions' are being tested*

	Smart Kalasatama	CANEMU Mynämäki
Developers	• Experimentation with new 'smart' ideas, gain experience with new digital solutions, learning for upscaling and diffusion from a large smart city site • Innovation platform and its business potential • Initially 'smart' digital solutions; energy and environment gained focus later	• Solutions for small municipalities to reduce CO2 emissions by 2030 • Early ideas focused on indigenous energy sources; later focus turned to energy efficiency, heat pumps and solar heat and power
Local officials and politicians	• New innovation and business ideas, environmental urban infrastructure, rapid experimentation, energy management solutions, business potential of new ideas	• Regional economic benefits, climate benefits, improved reputation, new competencies
Residents	• New, modern and cool city district close to the sea and city centre, technical solutions on smart energy use	• Practical energy solutions, more future-adapted ways of life
Funding bodies	• Experimentation with new business and technological ideas related to smart energy use, gain experience with new solutions, learning for upscaling and diffusion, business potential of new ideas	• Proof that small municipalities can find cost-effective ways to reduce CO2 emissions, proof of the plausibility of a green economy

being tested. Table 10.3 outlines some of the main viewpoints gathered in our cases in Smart Kalasatama and CANEMU Mynämäki.

This diversity is also reflected in the metrics that interviewees proposed for evaluation. For example, in Kalasatama, proposals ranged from the creation of new businesses and spin-offs to measures like carbon dioxide emissions per square meter. Similarly in CANEMU and Mynämäki, while solid metrics had been developed for monitoring carbon dioxide emission reductions, many wanted to see evidence of the impact of the experiment on the local economy.

The cases also reflect an *evolution of expectations*, where some original ideas might lose focus and some new ones might gain more attention as the experiment develops. Because of this, it might not always make sense to assess outcomes vis-à-vis the original expectations. For example, in the case of Kalasatama, energy and environmental objectives had gained focus after the city plan had been approved, which could create a problem in evaluation if new objectives are not accounted for. The new environmental focus raised expectations for more radical and ambitious low-carbon solutions, such as nearly zero-energy buildings, solar panels, green roofs and carless areas. Later, attention shifted towards building a sense of community and creating more sustainable and smart business opportunities. In Mynämäki, on the other hand, locals and some developers had originally envisaged the development of indigenous energy sources (rushes from the bay, sugar beet stalks), whereas over time the focus turned to more low-risk implementation of existing solutions such as energy-efficient HVAC, LEDs, heat pumps and solar heat and power. These developments were a result of learning processes in the course of experiments, but sometimes also contributed to a mismatch between expectations and outcomes, posing a problem for evaluation.

Local experiments, from an SNM perspective, are also expected to create *alignment of expectations*, so a constituency is formed to support the emerging solutions (Raven et al., 2008). We found indications of some alignment processes, particularly in Mynämäki, where the overall style of experimentation was more bottom-up and iterative, and on the other hand, there are stronger economic constraints in the realisation of the experiment. It is early to say, but according to our observations, we found less alignment in Kalasatama, which is a larger and more company-driven development, which explicitly aims to foster diverse (or even competing) product and service development processes. There were even signs of conflicting expectations and some disappointment of lower than expected ambition levels related to sustainability objectives.

10.4.2 Diverse, Fragmented Learning

In terms of types of lessons to be learned, local sustainability experiments are clearly more complex than envisaged in the basic SNM model of variation, selection and retention. While the SNM perspective recognises the evolution of the socio-technical system as one of the purposes of experimentation (e.g. Raven et al., 2008), original formulations focused on particular technologies, rather than the local people who might prefer an alternative technology (Schreuer, Ornetzeder and Rohracher, 2010). This was evident in the bottom-up project in Mynämäki, where initial ideas like local biomass were exchanged for more financially viable ones. In Kalasatama, as well, some of the technical solutions are still contested (e.g. solutions for transmitting data between appliances and the electric grid). There is also mounting pressure for more ambitious energy solutions in buildings. This does not mean to say that technology developers cannot learn from local sustainability experiments – indeed they can and have. However, due to the combination of several technologies, the overall or summary lessons (e.g. from the perspective of funding bodies) are not always straightforward, and it is not obvious what emerging 'niche community' (Geels and Raven, 2006) would aggregate such lessons.

The combination of actors and technologies, however, appears to offer ample opportunities for participants to gain new skills within the local context. For example, in the Mynämäki case, several participants mentioned as a key benefit the new kinds of capabilities required and developed by the experiment, such as skills in making an energy town plan or learning to use (and sometimes even teach others to use) new technologies. Similarly, by being confronted with each other, even participants with contradictory views and interests gained new ideas and orientations, as was the case for companies from different sectors forced to co-operate in the Smart Kalasatama experiment. This type of learning is not about 'what works where and when' but about 'what we can do'. In this way, learning in experiments also changes the identity of participants.

Transfer of lessons to other sites and *inspiration* were the main media for expansion of the experiments envisaged by the participants, including the funding bodies. Indeed, such expansion was observed in our cases: for example, some of the 'agile piloting' in Kalasatama had branched off to other cities, whereas in Mynämäki, a successful joint purchasing process for solar panels was both replicated at other locations and scaled up into a nationwide joint purchasing initiative for all Finnish municipalities, with additional features such as leasing. Different parties, thus, took away different lessons from the experiment, and it is not the results of the experiment as a whole, but diverse lessons learned from it that are embedded and scaled as trans-local learning (Geels and Raven, 2006). In many instances, some degree of success and positive attention in the original experimental site were crucial to induce new localities to adopt them.

In contrast, we did not really find schemes for aggregating lessons of 'what does not work', which some of the national authorities in our workshop felt was a gap in the scheme of 'experimental governance'. In our cases, even national funding bodies and policymakers did not evaluate the experiment as a whole, apart from particular aspects such as product development processes or carbon dioxide emissions. Moreover, they did not appear to approach success and failure symmetrically, since they also need positive references that show that sustainable solutions can be found. This type of proof and encouragement is needed not only for technical and administrative but also for political and psychological reasons (Heiskanen et al., 2015).

Our observations suggest that learning from failure may be impeded by the performative and political nature of local sustainability experiments. As Hodson and Marvin (2007) and Evans and Karvonen (2014) have shown, the management of expectations is crucial when knowledge production occurs in a highly public space, and hence, it may be difficult to acknowledge failure. In addition, if the experiment is realised as a co-operation between various actors, the future co-operation possibilities need to be kept open, which makes the systematic scrutiny of failures in experimentation outcomes problematic. While the notion of 'fail forward' might be acknowledged in principle, our cases show that there are many reasons why it might be more difficult to apply in practice. Our observations suggest that participatory evaluations should consider how success or failure of an experiment influences participants' identities. Hence, local sustainability experiments might not stand outside the participants to a sufficient extent to be evaluable separately from the achievements of the participants themselves. If we take seriously the performative and political nature of place-based experiments, then they reflect significantly on the worth of the place where the experiment is performed and on the worth of the people – in different roles – involved in its performance.

10.4.3 How Commitment Influences Outcomes

Dahlér-Larsen (2001) described as 'magic' those kinds of programmes that are able to create such a commitment that participants force even 'flawed' mechanisms to work. We are not suggesting that either Smart Kalasatama or CANEMU in Mynämäki are based on flawed ideas or principles. However, we did observe several features of commitment-creating magic at work. For example, in Kalasatama, stakeholders were impressed by the ability of the organisation in charge for creating a positive image of the area, which it did by organising events, visits and presentations. Thanks to rhetoric and drive, this organisation has painted a picture of a desired future towards which the experimentation in Kalasatama will lead the development. According to our interviewees, the organisation managing the project makes the innovations real even before the experiment or pilot is functioning by producing a story that encourages the developers and inhabitants. As one of our interviewees stated, 'there is a lot of activities and much visibility for the experiment even before the pilots were, or are, in action. The organisation creates a complete reality merely through the inspiring rhetoric.'

In Mynämäki, initial commitment was fairly limited, but it was greatly enhanced by a news item in a nationwide economics magazine stating that 'Change begins in Mynämäki'. Commitment was further enhanced by small wins, such as cost savings in public buildings, which were expected to 'reduce the tax burden of residents. While some had been initially concerned that the experiment might be foolish and rash, the 'sensible' outcomes and concrete benefits reduced their anxiety and created commitment to continue with the experiment and even raise its ambitions.

Dahlér-Larsen (2001) described as 'tragic' such programmes that undermine the commitment of the participants to the production of programme outcomes because the programme sends the participants a message that they are incompetent. We are not suggesting that there are any genuinely 'tragic' elements in either of the programmes. But we do suggest that an accumulation of failures – or even the delayed delivery of results – can lead to a loss of commitment that might reflect on even principally sound solutions and that, through such loss of commitment, learning from failures also remains limited.

10.5 Discussion and Implications

We have collected participants' perspectives on success and lessons learned in two different case studies of sustainability experimentation and reflected on how these contribute to learning and embedding new 'solutions' locally and in wider society. Following the realistic evaluation and theories of change literatures (Blamey and McKenzie, 2007), we tried to identify different participants' views on what 'solutions' the experiment is trialling, testing or promoting in the local 'context' and how one could judge if it has been successful. This revealed a diversity of expectations, as well as some degree of evolution and alignment. Our cases serve to illustrate how complex local experiments are to evaluate, given the diversity (and change) in the 'solutions', actors and interests involved.

In contrast to a randomised control trial testing an intervention, learning is only one of the aims of local sustainability experiments. Once locals are involved, the creation of a better living environment unavoidably becomes an aim of the experiment. As regards the embedding of new solutions locally and beyond the experimental site, we identified diverse forms of learning, which might or might not connect. These are schematically presented in Figure 10.1, and elaborated in the following discussion.

In our cases, the *local embedding of capabilities* appeared as a significant form of learning (Figure 10.1a). This includes the development of participant capabilities through learning by doing, or interacting and using (Jensen et al., 2007). It also includes and feeds into the development of new identities, where participants start to identify with the experiment or redefine their identities vis-à-vis other participants. Embedding also involves evolution and (partial) alignment of expectations, as well as development of commitment and confidence. Successful experiments can also 'radiate' confidence to other similar sites (see Raven et al., 2008) and to funding bodies, increasing confidence in society's ability to solve sustainability problems. In this kind of learning, it is not so much a question of 'what works' but rather an inspirational statement: 'We can do it!'

Distributed learning was, in our cases, the main form in which learning was transported beyond the original sites (Figure 10.1b). Participants representing different 'solutions' would take lessons and carry them to different sites to accumulate further niches and to inspire new participants. These sites might represent different combinations of 'solutions' and 'contexts', yet we found limited consideration of why these particular sites were chosen, apart from the fact that they provided opportunities and could draw on the original experiment as a 'reference'. In this form of evolutionary learning, negative lessons were usually not explicitly recorded or analysed, but rather accumulated in the tacit knowledge base of the actors involved.

Learning from multiple experiments appears to be problematic, according to our case studies (Figure 10.1c). Since expectations are diverse and evolve, different stakeholders can hold quite different views on success and failure; hence, the development of a comprehensive evaluation is a complex task. Among funding bodies, we found limited evaluation of the local sustainability experiments as a whole. We also found limited consideration of the lessons to be learned from combining particular 'solutions' in a particular 'context'. According to our data, at least in Finland, which has just launched its 'experimental culture', local sustainability experiments still appear to serve more as 'exemplars' and sources of inspiration than as a basis for analysis, reflection and consideration of 'what works where'.

This paper has only scratched the surface of evaluation in local sustainability experiments. We examine only two case studies in one country. Yet our observations concur with previous research (Hargreaves, 2011; Heiskanen et al., 2015; Späth and Rohracher, 2010; Kivimaa et al., 2017). As we illustrate, local sustainability experiments are often testing several things at once (Naess et al., 2012) in a

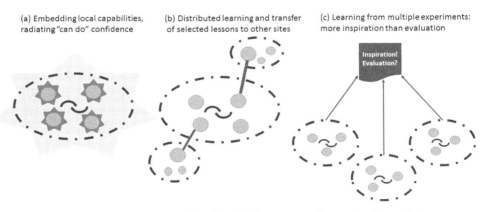

(a) Embedding local capabilities, radiating "can do" confidence

(b) Distributed learning and transfer of selected lessons to other sites

(c) Learning from multiple experiments: more inspiration than evaluation

Inspiration! Evaluation?

Figure 10.1. Forms of learning identified in the case studies of local sustainability experimentation.

multi-interest context of local politics (Quitzau et al., 2012). Until now, we have found limited attempts to aggregate lessons learned across experimental sites and the diverse solutions tested (and their interactions). This does not mean that there is no aggregate learning, but we suggest that the aggregate learning focuses on successful and inspirational outcomes, while less attention is paid to unsuccessful ones (see Kivimaa et al., 2017).

This is a sensitive topic for further evaluation, since local sustainability experiments are not only about learning in a technoscientific sense. They are also performative and political (Späth and Rohracher, 2010; Heiskanen et al., 2015). Participatory evaluation is the obvious choice for evaluating local sustainability experiments as a form of networked governance (Hertting and Vedung, 2012). Still, it might not necessarily be inconclusive about the worth of the particular solutions applied (even in their particular context). This is because judgements and expectations evolve in relation to what is being tried out and how it reflects on the worth of the participants themselves. Following Dahler-Larsen (2001), experiments modify contexts (indeed that is their purpose) and can create or undermine participants' commitment and confidence.

Given these conditions, evaluation of local sustainability experiments is extremely complex. While we feel that further research is crucial, we venture some tentative implications.

1. Local sustainability experiments are in dire need of evaluation if they are to become further embedded and to contribute to climate governance on a broader scale and in the long-term (including the formal state bureaucracy and governmental accountability).
2. Existing evaluation methods are inappropriate, since they fail to systematically account for the inextricability of the 'solutions' tested by the experiments and the 'contexts' in which they are tested. So, evaluation procedures need to not only incorporate but also go beyond realistic evaluation designs to capture emergent aims and the socially constructed outcomes of local climate experiments.
3. Evaluation should be participatory, but allow for diverse viewpoints and completely different understandings of what the experiment is about. Evaluation methods should be able to capture evolution in the aims of the experiment. They should also be able to deal with the fact that experiments may create or undermine commitment, thus obscuring the assessment of the validity of any 'mechanisms' tested. The challenge is to draw broader lessons while acknowledging the inseparability of 'mechanisms', contexts and participant identities.
4. 'What works when and where' might not be the main question local sustainability experiments aim to address. Other questions, such as seeking to evaluate

the scope for distributed learning, inspiration and development of commitment and confidence, might be equally valuable. Evaluation methods should be able to capture the inspirational and motivational aspects of local sustainability experiments, as well as the opportunities provided for various 'niche' solutions, but there should also be means to capture and understand aspects that did not work well, and why.

References

Argyris, C. (2003). A life full of learning. *Organization Studies*, 24(7), 1178–1193.

Blamey, A., and Mackenzie, M. (2007). Theories of change and realistic evaluation peas in a pod or apples and oranges? *Evaluation*, 13(4), 439–455.

Blok, A. (2013). Urban green assemblages. *Science & Technology Studies*, 26(1), 5–24.

Bulkeley, H., and Castán Broto, V. (2013). Government by experiment? Global cities and the governing of climate change. *Transactions of the Institute of British Geographers*, 38(3), 361–375.

Dahler-Larsen, P. (2001). From programme theory to constructivism on tragic, magic and competing programmes. *Evaluation*, 7(3), 331–349.

Dahler-Larsen, P., and Schwandt, T. A. (2012). Political culture as context for evaluation. *New Directions for Evaluation*, 2012(135), 75–87.

De Búrca, G., Keohane, R. O., and Sabel, C. (2014). Global experimentalist governance. *British Journal of Political Science*, 44(3), 477–486.

De Schutter, O. (2010). The role of evaluation in experimentalist governance: Learning by monitoring in the establishment of the area of freedom, security, and justice. In Sabel, C. F., and Zeitlin, J. (eds.), *Experimentalist Governance in the European Union: Towards a New Architecture*. Oxford: Oxford University Press, 261–296.

Evans, J., and Karvonen, A. (2014). 'Give me a laboratory and I will lower your carbon footprint!' Urban laboratories and the governance of low-carbon futures. *International Journal of Urban and Regional Research*, 38(2), 413–430.

Evans, J. P. (2011). Resilience, ecology and adaptation in the experimental city. *Transactions of the Institute of British Geographers*, 36(2), 223–237.

Friedman, V. J. (2001). Designed blindness: An action science perspective on program theory evaluation. *American Journal of Evaluation*, 22(2), 161–181.

Geels, F. W., and Deuten, J. J. (2006). Local and global dynamics in technological development: A socio-cognitive perspective on knowledge flows and lessons from reinforced concrete. *Science and Public Policy*, 33(4), 265–275.

Geels, F., and Raven, R. (2006). Non-linearity and expectations in niche-development trajectories: Ups and downs in Dutch biogas development (1973–2003). *Technology Analysis & Strategic Management*, 18(3–4), 375–392.

Government Programme (2015). Programme of Prime Minister Sipilä's government. Accessed Oct 3rd 2017, http://valtioneuvosto.fi/sipilan-hallitus/hallitusohjelma?p_p_id=56_INSTANCE_SSKDNE5ODInk&p_p_lifecycle=0&p_p_state=normal&p_p_mode=view&p_p_col_id=column-2&p_p_col_count=1&_56_INSTANCE_SSKDNE5ODInk_languageId=en_US

Hargreaves, T. (2011). What lessons get shared? Case studies of community energy. Grassroots innovation research briefing 9. Accessed Oct 3rd 2017, https://grassrootsinnovations.files.wordpress.com/2012/04/gi-9-cise-3rd-party-case-studies.pdf

Heiskanen, E., Jalas, M., Rinkinen, J., and Tainio, P. (2015). The local community as a 'low-carbon lab': Promises and perils. *Environmental Innovation and Societal Transitions*, 14, 149–164.

Hertting, N., and Vedung, E. (2012). Purposes and criteria in network governance evaluation: How far does standard evaluation vocabulary takes us? *Evaluation*, 18(1), 27–46.

Hodson, M., and Marvin, S. (2007). Cities mediating technological transitions: The adaptability of infrastructure and infrastructures of adaptability. *International Handbook of Urban Policy*, Cheltenham, Edward Elgar, 240–258.

Hodson, M., and Marvin, S. (2009). Cities mediating technological transitions: Understanding visions, intermediation and consequences. *Technology Analysis & Strategic Management*, 21(4), 515–534.

Hoffmann, M. J. (2011). *Climate Governance at the Crossroads: Experimenting with a Global Response after Kyoto*. Oxford, Oxford University Press.

Holm, J., Stauning, I., and Søndergård, B. (2011). Local climate mitigation and eco-efforts in housing and construction as transition places. *Environmental Policy and Governance*, 21(3), 183–198.

Hoogma, R., Weber, M., and Elzen, B. (2005). Integrated long-term strategies to induce regime shifts towards sustainability: The approach of strategic niche management. In Weber, M. and Hemmelskamp, J. (Eds.) *Towards Environmental Innovation Systems*. Berlin Heidelberg: Springer, 209–236.

Jensen, M. B., Johnson, B., Lorenz, E., and Lundvall, B. Å. (2007). Forms of knowledge and modes of innovation. *Research Policy,* 36(5), 680–693.

Jordan, A., and Huitema, D. (2014). Innovations in climate policy: The politics of invention, diffusion, and evaluation. *Environmental Politics*, 23(5), 715–734.

Kemp, R., Schot, J., and Hoogma, R. (1998). Regime shifts to sustainability through processes of niche formation: The approach of strategic niche management. *Technology Analysis & Strategic Management*, 10(2), 175–198.

Kivimaa, P., Hildén, M., Huitema, D., Jordan, A., and Newig, J. (2017). Experiments in climate governance – a systematic review of research on energy and built environment transitions. *Journal of Cleaner Production,* 169, 17–29.

Mickwitz, P., Hildén, M., Seppälä, J., and Melanen, M. (2011). Sustainability through system transformation: Lessons from Finnish efforts. *Journal of Cleaner Production*, 19(16), 1779–1787.

Mulgan, G. (2014). *Innovation in the Public Sector: How Can Public Organisations better Create, Improve and Adapt?* London: Nesta.

Naess, P., and Vogel, N. (2012). Sustainable urban development and the multi-level transition perspective. *Environmental Innovation and Societal Transitions,* 4, 36–50.

Overdevest, C., Bleicher, A., and Gross, M. (2010). The experimental turn in environmental sociology: Pragmatism and new forms of governance. In M. Gross and H. Heinrichs (Eds.). *Environmental Sociology*. The Netherlands: Springer Netherlands, 279–294.

Pawson, R., Tilley, N. (1997). *Realistic Evaluation*. London: Sage.

Pearce, W., and Raman, S. (2014). The new randomised controlled trials (RCT) movement in public policy: Challenges of epistemic governance. *Policy Sciences*, 47(4), 387–402.

Quitzau, M.-B., Hoffmann, B., and Elle, M. (2012). Local niche planning and its strategic implications for implementation of energy efficient technology. *Technological Forecasting & Social Change*, 79, 1049–1058.

Raven, R., Kern, F., Smith, A., Jacobsson, S., and Verhees, B. (2016). The politics of innovation spaces for low-carbon energy: Introduction to the special issue. *Environmental Innovation and Societal Transitions*, 18, 101–110.

Raven, R. P., Heiskanen, E., Lovio, R., Hodson, M., and Brohmann, B. (2008). The contribution of local experiments and negotiation processes to field-level learning in emerging (niche) technologies meta-analysis of 27 new energy projects in Europe. *Bulletin of Science, Technology & Society*, 28(6), 464–477.

Rohracher, H. (2008). Energy systems in transition: Contributions from social sciences. *International Journal of Environmental Technology and Management*, 9(2–3), 144–161.

Sabel, C. F., and Zeitlin, J. (2008). Learning from difference: The new architecture of experimentalist governance in the EU. *European Law Journal*, 14(3), 271–327.

Schot, J., and Geels, F. W. (2008). Strategic niche management and sustainable innovation journeys: Theory, findings, research agenda, and policy. *Technology Analysis & Strategic Management*, 20(5), 537–554.

Schreuer, A., Ornetzeder, M., and Rohracher, H. (2010). Negotiating the local embedding of socio-technical experiments: A case study in fuel cell technology. *Technology Analysis & Strategic Management*, 22, 729–743.

Späth, P., and Rohracher, H. (2010). 'Energy regions': The transformative power of regional discourses on socio-technical futures. *Research Policy*, 39, 449–458.

Vedung, E. (1997). *Public Policy and Program Evaluation*. New Brunswick, NJ: Transaction Publishers.

Vedung, E. (2006). Evaluation research. In Peters, B.G., and Pierre, J. (eds.), *Handbook of Public Policy*. London, Sage Publications, 397-416.

Voytenko, Y., McCormick, K., Evans, J., and Schliwa, G. (2016). Urban living labs for sustainability and low carbon cities in Europe: Towards a research agenda. *Journal of Cleaner Production*, 123, 45–54.

11

The City of Permanent Experiments?

ANDREW KARVONEN

11.1 Introduction

Over the last decade, a multitude of experiments has emerged to catalyse climate change transitions (Evans, 2011; Castán Broto and Bulkeley, 2013; Evans and Karvonen, 2014; Bulkeley, Castán Broto and Edwards, 2015). These experiments involve activities that go beyond traditional state activities of policymaking, regulation, incentives and collective service provision (Biermann and Pattberg, 2008; Okereke, Bulkeley and Schroeder, 2009; Hoffmann, 2011). This is evident in cities where climate change experiments feed on innovation and creativity 'to reinvigorate and stretch traditional avenues of governance to face the multifaceted challenges posed by a changing climate' (Karvonen, Evans and van Heur, 2014: 110). In effect, they provide a situated and real-world evidence base for how a low-carbon world could be realised, and they have the potential to fundamentally change the way that cities are conceived, built and managed.

The majority of urban climate change experiments are designed to be geographically and temporally bounded. This suggests that 'experiments are critical *sites* through which visions of low carbon cities are created, networks built and learning enacted' (Bulkeley and Castán Broto, 2013: 373; emphasis added). Experimental setups are constructed and operated in inscribed spaces that are variously labelled as urban laboratories, innovation districts, demonstration projects and test beds (Karvonen and van Heur, 2014; Marvin and Silver, 2016). Restricting the spatial and temporal reach of climate change experiments is useful to accelerate innovation activities and to realise actual changes on the ground (van Buuren et al., Chapter 8). The outcomes of these experimental activities can then be assessed against their intended outcomes through processes of formal evaluation (Heiskanen and Matchoss, Chapter 10). If deemed successful, the findings can be upscaled and transferred to other locales through policymaking, development and management activities. And if deemed unsuccessful, the experiments can be revised and conducted anew or simply abandoned (van Buuren et al., Chapter 8). Thus, experiments serve as a tangible way to institutionalise and formalise climate change governance (Angelouvski and Carmin, 2011; Bulkeley and Castán Broto, 2013).

But what if urban experiments did not scale up or travel to other locales? What if, instead of informing existing modes of urban governance, they became the dominant approach to governing cities? What would a 'city of permanent experiments' look like and how would it function? This chapter explores the implications of experimentation as the new mode of governance for twenty-first century cities. Specifically, it is argued here that experiments might not simply serve as one-off trials to provide evidence and justification for new low-carbon policies, regulations and service provision through existing circuits of policymaking and regulation. Instead, these activities are emerging as a new mode of governance in themselves (Sabel and Zeitlin, 2008, 2010; Bulkeley et al., 2015). Castán Broto and Bulkeley (2013: 1947) argue that experiments 'serve as potentially powerful means to create new forms of governing everyday life and city circulations, simultaneously reinforcing and reinventing the landscape of governance'.

By moving away from notions of upscaling and transfer of experimental outcomes, this chapter points towards an emerging form of urban governance characterised by the embrace of uncertainty and contingency, the fostering of recursive learning processes and the increasing spatial fragmentation of cities into experimental districts. This speculative inquiry is not intended as a forecast or prognostication. Instead, it is an attempt to reflect on the larger meaning of experimental practices and where these activities could potentially lead if they continue to propagate and become the dominant mode of climate governance. Of particular interest are the qualities of experiments and how they challenge and will potentially alter the future trajectory of cities in both intended and unintended ways. Such an emerging 'city of permanent experiments' has a number of significant implications on the politics of twenty-first-century urban development.

11.2 The Allure of Urban Experimentation

Experimentation is an enticing twenty-first-century response to a range of contemporary crises, ranging from climate change and global financial collapse to widespread social unrest and religious extremism. The appeal of experimentation lies in its promise to introduce new conceptions and configurations to the evolution of cities (Karvonen and van Heur, 2014). With respect to climate change, a wide range of public, private and third-sector actors are developing and conducting interventions to alter the socio-technical function of energy, transport, food, manufacturing, healthcare, buildings, finance and so on (Bulkeley and Castán

Broto, 2013; Bulkeley et al., 2015). As Evans, Karvonen and Raven (2016: 2) note, 'Smart cities, eco-cities, low carbon urbanism, urban living labs, happy cities and sustainable urban development all draw on the idea that experimentation can generate more liveable, prosperous and sustainable urban futures.' Urban experiments embody both top-down and bottom-up forms of organisation that involve multiple stakeholders with varying aims and objectives. A common thread of these activities is to provide visible and tangible interventions that demonstrate that action is being taken. Bulkeley and colleagues (2015: 50) note that 'making experiments is an attempt to negotiate the perceived scale divide between the unfolding of the problem and the possibilities of intervention'. These are normative and highly visible activities that promise improved urban futures.

The rising popularity of experimentation in recent years has led to the term being used as a synonym for almost any form of change or deviance from 'business as usual'. Gross (2010a: 66) notes that 'the concept of experimentation comes to have the same meaning as development, complexity, interconnection, globalization and so comes to mean the same as virtually anything that is subject to change'. To counter this trend, it is important to understand urban experimentation not simply as a set of ad hoc or novel activities that diverge from the status quo but also as situated and purposive modes of trialling new configurations of socio-technical systems to develop knowledge to respond to a particular problem. Karvonen and van Heur (2014: 383) define urban experimentation as: '1) involving a specific set-up of instruments and people that 2) aims for the controlled inducement of changes and 3) the measurement of these changes'. This more formalised definition of experimentation highlights the simultaneous actions of observation and intervention with a specific end goal in mind. Robust urban experiments follow the logic of the scientific method and are carefully planned and executed. They have an explicit aim of measuring the impact of the intervention and comparing this to existing or baseline conditions. This process of experimentation creates learning loops to inform new modes of climate governance (Collins and Ison, 2009; Evans and Karvonen, 2011).

In addition, it is commonly assumed that the knowledge generated from experiments is intended to be scaled up both temporally and spatially (Brown and Vergragt, 2008; Evans, 2011; Nevens et al., 2013; Bulkeley et al., 2015). Evans and colleagues (2016: 4) argue that 'part of the allure of experimentation is based on the assumption that it is possible to scale up from an individual project to the city through a process of trialling, learning and rolling out'. Thus, it is the outcomes of these experiments that have the greatest importance as they are translated into policy and regulatory frameworks and other existing modes of governance. But what if experiments, rather than informing existing modes of urban governance, were an end in themselves? What if experiments embodied a new form of urban governance? Bulkeley and colleagues (2015: 243) argue that 'rather than being a matter of scaling up experimentation, its transformative potential comes through the traction generated by refashioning existing socio-technical configurations by translating between climate governmentalities and existing forms of social and material order'. It is here that experimentation emerges as a new authority for climate change mitigation and adaptation activities. Experiments become embedded in climate change activities not through scaling and transfer but through the gradual replacement of existing modes of governance. This points towards an emerging 'city of permanent experiments' with significant implications on how urban development unfolds.

The following sections consider three potential achievements of experimental governance: (1) the embrace of uncertainty and contingency, (2) the generation of recursive learning loops and (3) the fragmentation of the city into districts of innovation. Taken as a whole, these qualities of urban experiments suggest a new epistemology of urban governance that emphasises situated knowledge generation and action with unclear outcomes.

11.3 Embracing Uncertainty

Uncertainty is a defining characteristic of climate governance in cities, both in defining the problem at hand as well as devising potential solutions. The scale and magnitude of climate change creates multiple contingencies that are simultaneously physical, social, political and economic. Urban experiments provide a mechanism to embrace ignorance, surprise and the unknown (Callon, Lascoumes and Barthe, 2009; Gross, 2010b; Kullman, 2013). As Karvonen and van Heur (2014: 387) note, 'real-world experimentation is founded on the idea that one is compelled to act despite uncertainties and gaps in knowledge.' Uncertainty acknowledges the provisional and dynamic character of urban evolution (Simone, 2013; Karvonen et al., 2014; Silver, 2014). Unlike conventional urban governance, experiments are open to and feed on the messy and contingent character of policymaking (Meadowcroft, 1999, 2009). It is understood that this mess is where new low-carbon futures can be forged.

Such a perspective resonates with a range of contemporary theoretical debates including post-structuralism, pragmatism, Mode 2 Science, transdisciplinarity, applied innovation and knowledge co-production (e.g. Gibbons et al., 1994; Nowotny, Scott and Gibbons, 2001; Ramadier, 2004; Benneworth, Charles and Madanipour, 2010; Evans and Karvonen, 2011). As Karvonen and colleagues (2014: 106) argue, 'The appeal of experimentation lies in its ability to harness *radical contingency* in the service of urban development' (emphasis in original). Uncertainty then becomes a catalyst for action rather than a hindrance to urban development. Further, it embodies a 'particular style of adaptive governance' (Evans, 2011: 255) that responds to 'the emergent properties of the system' (Bulkeley and Castán Broto, 2013: 366). Experimentation replaces the procedural character of urban policymaking in favour of agile and responsive interventions. This recasts climate governance as edgy and dynamic, acknowledging and embracing the risks and unknowns of urban development processes.

Of course, uncertainty is not restricted to urban climate change experiments. The last decade has witnessed the rise of precarious urban conditions involving temporary land use, the appropriation of vacant buildings, pop-up trucks and stores, energy and water precarities and related dialogues (e.g. Neilsen and Rossiter, 2008; Bishop and Williams, 2012; Andres, 2013; Tonkiss, 2013; Németh and Langhorst, 2014; Harris, 2015). The world as a whole seems to be moving towards more precarious states of existence as the durable edifices of twentieth-century liberal democracy slowly crumble. However, experimentation contrasts with these other activities by championing instability and unpredictability as a constructive, productive characteristic to reconfigure socio-technical systems. The unknown is an opportunity to develop new ideas and approaches. There are strong parallels here with well-known alternative eco-settlements such as Arcosanti (United Sates), Auroville (India), Findhorn (Scotland) and ZEGG (Germany) (see Miles, 2007; Pickerill and Maxey, 2009; Vanolo, 2013; Pickerill, 2016).

Christiania in Denmark is perhaps the most famous of these alternative eco-settlements. Initiated in 1971 on thirty-four hectares in Greater Copenhagen as a hippy community, the neighbourhood has evolved as a centre for social activism and lifestyles to counter mainstream capitalist society (Conroy, 1994; Vanolo, 2013). The urban settlement consists of about 400 buildings and around 1,000 residents and supports social services such as healthcare, schools and a post office as well as workshops for light industry, restaurants and bars. The settlement follows some local and national laws and regulations but much of day-to-day life is shaped through self-governance and participatory democracy. Vanolo (2013: 1789) notes that 'the internal governance structure of Christiania is inspired by ideas of collective self-government and direct democracy, so that relevant decisions are always taken by consensus and after protracted discussion and negotiation in thematic meetings.' It is this emphasis on self-governance that informs the settlement's character of precarity and fluidity. The status of Christiania (and the other eco-settlements noted previously) is constantly being negotiated and reworked to reflect the needs and desires of the residents.

In many ways, urban climate change experiments bring the ethos if not the politics of eco-settlements to conventional urban development; they reinforce uncertainty as an unavoidable and generative characteristic of the urban experience. A city of permanent experiments is thus a provisional achievement and one that is always 'in the making'. This also connects to recent theoretical contributions on assemblage theory and socio-technical urban systems (Anderson and McFarlane, 2011; Brenner, Madden and Wachsmuth, 2011; Farías, 2011; McFarlane, 2011b; Gopakumar, 2014) that recognise that cities do not have an end point, but are always unfolding and evolving – they are processes rather than products. As McFarlane (2011b: 650) notes, the city here is understood as a 'gathering process' rather than an end state. This fluidity of urbanisation involves the continual reassembly of socio-technical metabolisms and the city of permanent experiments recognises governance as an activity of continually negotiating and responding to contemporary pressures and opportunities. However, this embrace of fluidity raises significant questions about the agency of various urban stakeholders. What competencies and skills are required to nurture this ever-changing state of the urban? How is uncertainty put to work to realise alternative futures? Who is held accountable and how?

11.4 Developing Learning Loops

The embrace of uncertainty suggests that a city of permanent experiments has particular implications on flows of knowledge. Evans and Karvonen (2014: 426) note that experimentation involves 'empiricising the urban landscape through monitoring and instrumentation, and then materializing these empirics by feeding them into subsequent planning policy that will shape urban development'. The measurement and assessment of collected data provides situated, real-world feedback about the performance of an intervention for transport, energy, water and so on. This involves the development of reflexive governance approaches that are informed by monitoring and assessment. One of the principal benefits of urban climate change experiments is that they instil recursive learning into the governance of cities. Karvonen et al. (2014: 105) argue that 'experiments suggest recursive learning as a key component of enacting different urban futures.' If, as McFarlane (2011a: 360) argues, the 'city is a machine for learning', then experiments serve as the metaphorical sparkplug to generate an evidence base to inform decisions about urban development. This builds on notions of 'reflexive modernity' (Beck, 1995) to realise formal modes of learning and response that can catalyse urban transformations.

With experiments, measurement and assessment is formalised as a process to support or contest various knowledge claims and discursive positions about the future of cities (Bulkeley and Castán Broto, 2013). This provides new insights about cities that can be used to steer urban development (Karvonen et al., 2014). The empirical character of urban climate change experiments feeds into evidence-based policy activities where the governance of the city is informed by a robust base of findings rather than ideology (Sanderson, 2002). This is commonplace in the justification for smart urban development projects. Ubiquitous sensor networks are proposed to provide massive datasets to produce new understandings of urban function. These datasets provide an instantaneous as well as long-term evidence base while creating a continuous process of learning and feedback to optimise and manage the socio-technical metabolism of cities.

A contemporary example of the creation of continuous feedback loops is the futuristic smart city development of Songdo in South Korea (Lindsay, 2010; Halpern et al., 2013; Kshetri, Alcantara and Park, 2014; Carvalho, 2015). Songdo was initiated by the national government as a new development in the early 2000s as a 600-hectare international business district located 65 kilometres southwest of Seoul. The project is distinctive because it is an early example of applying a ubiquitous sensor network in a city. The local authority, in collaboration with Cisco, installed thousands of sensors in the built environment that feed data into a control centre. Technicians can control energy, transportation, water and other collective services in real time, creating a form of responsive urban governance. Reflecting on Songdo, Halpern and colleagues (2013) describe an emergent form of 'test-bed urbanism' where data collection, analysis and response are completely automated. This points towards a new epistemology of the city where 'self organization is its dominant rationality' (Halpern et al., 2013: 291). Self-regulation and self-adjustment become the end goal of urban learning loops, creating a city that is managed and steered by continuous short-term feedback rather than prescribed long-term planning. The roll-out of fully censored and algorithmic cities is several decades away but the possibilities of data-led urban development are beginning to emerge in various smart cities and point to how recursive learning might become an automated process in the future.

The emergence of learning loops again raises a number of questions. Who is involved in collecting and interpreting the data? What is done with the results? Does this reinforce the primacy of the expert as the ultimate arbiter of urban environmental governance? Are there ways to make learning loops more democratic and transparent?

11.5 Fragmenting the City

In addition to creating learning loops and embracing uncertainty, the city of permanent experiments has important spatial implications. A small number of urban experiments encompass an entire city or region (e.g. Masdar City and Songdo), but most are conducted at a more limited scale to accelerate innovation and delimit risk. Karvonen and colleagues (2014: 104–5) argue that 'the allure of the experiment lies in its ability to be radical in ambition while being limited in scope.' Designated spaces of innovation include urban and living laboratories, low-carbon districts, test beds, innovation zones, special purpose zones and demonstration sites (Graham and Marvin, 2001; Karvonen and van Heur, 2014; Evans et al., 2015; Voytenko et al., 2016). The emphasis on districts focuses on the unique socio-technical configurations of a particular context and provides real-world sites of demonstration that can be visited and experienced first-hand (Shapin and Shaffer, 1985; Reno, 2011; Späth and Rohracher, 2012).

This emphasis on districts as the principal spatial unit of experiments has important implications on climate change governance. Cities here are defined not as holistic entities but as composites or agglomerations of bounded spaces defined by the extent of their experimental reach. This resonates with interpretations of cities as a collection of quarters, neighbourhoods or villages (Neal, 2003; Bell and Jayne, 2004; Roodhouse, 2010). However, rather than creating a tidy and rational agglomeration of districts akin to a patchwork quilt, urban climate change experiments are more likely to create a landscape of overlaps and gaps. This is because their boundaries are defined by knowledge-generation activities rather than political, physical or social attributes. Within the city of permanent experiments, there is no conscious attempt to knit the city together into a seamless whole. The challenge of urban (climate) governance then involves managing or orchestrating the experiments within the districts as well as negotiating the boundaries, overlaps and gaps between districts (see Hölscher, Frantzeskaki and Loorbach, Chapter 7). Steering urban development involves engaging with the messiness of multiple experiments in an attempt to harmonise these activities. This raises significant issues about those existing urban service networks that transcend experimental boundaries (mass transit systems and roads, water and sewer networks, electricity supply, etc.) and their connection to experimental activities.

There is an intriguing historical parallel here with the rise of privatised collective service provisions in the nineteenth century when entrepreneurs introduced competing infrastructure networks of water, sewer and electricity. These competing service networks were eventually replaced by natural monopolies that were often publicly owned and managed to realise the modern infrastructural ideal (Graham and Marvin, 2001). With urban climate change experiments, there is the potential to return to the Wild West of urban service provision through the creation of messier 'post-networked' cities of the twenty-first century (Coutard and Rutherford, 2011, 2016). The emphasis on districts defined by experimental reach has the potential to increase social polarisation and economic disparity by favouring some districts with greater potential to address climate change mitigation and adaptation while neglecting others. And it also points to greater challenges in conceptualising and understanding cities as layered, fragmented, overlapping and messy.

More optimistically, the downscaling to districts hints at the potential for more face-to-face deliberation over the day-to-day and long-term functioning of cities.

Innovation districts could potentially foster more familiarity and interaction between stakeholders through participation in geographically bounded interventions. Moreover, this could bring climate change 'closer to home' by making experiments a normalised part of daily urban life (Castán Broto and Bulkeley, 2013). Experimentation is not an activity that is applied to urban residents but is something achieved through their input. This bottom-up form of urban development is personified in grassroots experiments, co-housing projects, transition towns, community energy initiatives and related movements (Middlemiss and Parrish, 2010; Chatterton, 2013, 2016; Feola and Nunes, 2014). The aim is to leverage communities of residents to effect change and, in the process, create a new polity (Walker, 2011; Karvonen, 2016).

A contemporary example of this is Ashton Hayes, a village of 900 people in Northwest England. In 2005, the villagers developed a bottom-up approach to climate governance to realise a shared goal of becoming England's first carbon-neutral community (Alexander, Hope and Degg, 2007; AHGCN, 2017). Eschewing input from their national and local governments, many (but not all) of the residents participate in collective efforts to reduce their carbon emissions through programmes of behavioural change, the establishment of a community energy company and the fostering of local business development. Alexander and colleagues (2007: 71) conclude that 'the carbon neutral project has enhanced the capacity of the Ashton Hayes community for self-management and self-reliance.' Of particular importance are the door-to-door surveys conducted by volunteers on a periodic basis to monitor carbon-reduction progress over time. Examples such as Ashton Hayes suggest that the city of permanent experiments has the potential to downscale climate change activities and, in the process, create spaces for more participatory forms of political deliberation and action. However, these grassroots activities do not emerge on their own. They require leadership and organisation just like any other form of governance; the major difference here is that it comes from the residents rather than from the local government.

A number of questions arise with the reconfiguration of cities into experimental districts. How do the districts connect to one another? Is there a role for local and regional government bodies? What is to be done with districts where no experiments are being undertaken or where these is no polity to carry them out? How are infrastructure networks that bridge multiple districts governed?

11.6 The City of Permanent Experiments

The twentieth-century era of city building is famously summarised in Daniel Burnham's proclamation, 'Make no little plans; they have no magic to stir men's blood' (see Gleeson, 2012). While there continue to be ambitious and high-profile projects to realise the twenty-first-century city, the emergence of urban experiments points towards small-scale, tentative interventions that incrementally shape urban development. In effect, experiments are the 'little plans' that Burnham warned us about over a century ago. This counters the reliance of conventional urban development on ambitious, comprehensive and long-term planning. Instead, reinventing cities involves the embrace of uncertainty, the development of learning loops and a focus on districts. Where it is typically assumed that climate change experiments inform and influence existing urban governance processes through processes of scaling and transfer, the city of permanent experiments suggests that experimentation is an end in itself. Bulkeley and Castán Broto (2013: 363) argue that 'such interventions are a critical means through which governing as normal takes place.' This embodies new forms of governance that are reflexive, responsive and spatially delineated.

This 'institutionalisation' of experimentation as a form of reflexive governance has significant political implications (Hoffman, 2011; Bulkeley and Castán Broto, 2013; Karvonen et al., 2014; Bulkeley et al., 2015; Evans et al., 2016). This emerging form of politics is neither inherently progressive nor regressive. While there is great potential for experiments to provide 'grist in the urban mill, creating conflict, sparking controversy, offering the basis for contested new regimes of practice' (Bulkeley and Castán Broto, 2013: 367–368), these activities can also be used to reproduce and extend existing modes of neo-liberal and capitalist modes of urban development. The politics of climate change experiments are ultimately determined by the context of these activities, the range of stakeholders involved and the ways in which the experiments are carried out.

Of particular interest to the politics of urban climate change experiments is the potential for experiments to enhance democratic participation in urban governance. The focus on the district scale suggests the opportunity to develop more discursive and participatory forms of governance through situated activities with multiple stakeholders. However, it is important to question who is allowed to take part in both the institutional and practical activities of experimentation (Meadowcroft, 2009; Evans et al., 2016). The rise of co-production as a driver for urban socio-technical transitions, as featured in urban living labs, emphasises the need to include all stakeholders in decisions about how cities evolve and change over time. As Evans and colleagues (2015: 1) write, '[c]o-production involves stakeholders understanding each others' contexts, in order to work together to frame research that delivers more effective solutions.' This suggests that experimentation has the potential to benefit from collective urban governance that brings together multiple voices (Voytenko et al., 2016) while also addressing the democratic deficit of contemporary urban politics.

On the other hand, these activities of co-production could simply validate predetermined agendas under the guise of democratic deliberation (Laurent,

2011). Thus, it is unclear if experiments provide a platform for controversies and differences to be deliberated and debated or if they are simply a means to reinforce the status quo (Karvonen et al., 2014). This suggests that there is a need to shift our collective gaze away from experimental activities and towards the stakeholders who are designing and conducting the experiments as well as those individuals and organisations that are using the experimental outcomes to affect change. It is through such a reflective and expansive consideration of urban climate change experiments as a political endeavour that we can assess their broader implications on urban development.

Ultimately, it is important to pay attention to this emerging agenda of urban climate change experimentation because it has the potential to transform urban governance and politics in ways that are yet to be determined. Reflecting on the ubiquitous sensing agenda in Songdo, Halpern and colleagues (2013: 300) ominously conclude that 'these experiments will concern us long before their outcome is clear.' Whether we are shortening the time frames of urban planning, creating feedback loops of learning, providing an arena for inclusive deliberation and decision-making or reinforcing existing trends of neo-liberal capitalism and globalisation, experimentation appeals to a wide range of urban stakeholders for disparate (and sometimes contradictory) reasons. These activities introduce particular modes of governance with significant long-term implications (McFarlane, 2011b; Bulkeley and Castán Broto, 2013). A fundamental challenge of experimentation is 'to bring people and infrastructures together in ways that are capable of realising significantly different and more sustainable urban futures in socially just and democratic ways' (Evans et al., 2016: 10). As such, urban climate change experiments are not simply a curiosity or a fleeting moment in the history of cities but provide a glimpse of the future of urban development.

References

Alexander, R., Hope, M., and Degg, M. (2007). Mainstreaming sustainable development – a case study: Ashton Hayes is going carbon neutral. *Local Economy*, 22, 62–74.

Anderson, B., and McFarlane, C. (2011). Assemblage and geography. *Area*, 43, 124–127.

Andres, L. (2013). Differential spaces, power hierarchy and collaborative planning: A critique of the role of temporary uses in shaping and making places. *Urban Studies*, 50, 759–775.

Angelouvski, I., and Carmin. J. (2011). Something borrowed, everything new: Innovation and institutionalization in urban climate governance. *Current Opinion in Environmental Sustainability*, 3, 169–175.

Ashton Hayes Going Carbon Neutral (AHGCN) (2017). Ashton Hayes Going Carbon Neutral. Access 6 February 2017. www.goingcarbonneutral.co.uk

Beck, U. (1995). *Ecological Enlightenment: Essays on the Politics of the Risk Society*. Atlantic Highlands, NJ: Humanities Press.

Bell, D., and Jayne, M. (eds.) (2004). *City of Quarters: Urban Villages in the Contemporary City*. London: Ashgate.

Benneworth, P., Charles, D., and Madanipour, A. (2010). Building localized interactions between universities and cities through university spatial development. *European Planning Studies*, 18, 1611–1629.

Biermann, F., and Pattberg, P. (2008). Global environmental governance: Taking stock, moving forward. *Annual Review of Environment and Resources*, 33, 277–294.

Bishop, P., and Williams, L. (2012). *The Temporary City*. London: Routledge.

Brenner, N., Madden, D. J., and Wachsmuth, D. (2011). Assemblage urbanism and the challenges of critical urban theory. *City*, 15, 225–240.

Brown, H. S., and Vergragt. P. J. (2008). Bounded socio-technical experiments as agents of systemic change: The case of a zero-energy residential building. *Technological Forecasting and Social Change*, 75, 107–130.

Bulkeley, H., and Castán Broto, V. (2013). Government by experiment? Global cities and the governing of climate change. *Transactions of the Institute of British Geographers*, 38, 361–375.

Bulkeley, H., Castán Broto, V., and Edwards, G. A. S. (2015). *An Urban Politics of Climate Change: Experimentation and the Governing of Socio-Technical Transitions*. London: Routledge.

Callon, M., Lascoumes, P., and Barthe, Y. (2009). *Acting in an Uncertain World: An Essay on Technical Democracy*. London: MIT Press.

Carvalho, L. (2015). Smart cities from scratch? A socio-technical perspective. *Cambridge Journal of Regions, Economy and Society*, 8, 43–60.

Castán Broto, V., and Bulkeley, H. (2013). Maintaining climate change experiments: Urban political ecology and the everyday reconfiguration of urban infrastructure. *International Journal of Urban and Regional Research*, 37, 1934–1948.

Chatterton, P. (2013). Towards an agenda for post-carbon cities: Lessons from Lilac, the UK's first ecological, affordable cohousing community. *International Journal of Urban and Regional Research*, 37, 1654–1674.

Chatterton, P. (2016). Building transitions to post-capitalist urban commons. *Transactions of the Institute of British Geographers*, 41, 403–415.

Collins, K., and Ison, R. (2009). Jumping off Arnstein's ladder: Social learning as a new policy paradigm for climate change adaptation. *Environmental Policy and Governance*, 19, 358–373.

Conroy, A. (1994). *Christiania: The Evolution of a Commune*. Amsterdam: International Institute of Social History.

Coutard, O., and Rutherford, J. (2011). The rise of post-networked cities in Europe? Recombining infrastructural, ecological and urban transformations in low carbon transitions. In Bulkeley, H., Castán Broto, V., Hodson, M., and Marvin, S. (eds.), *Cities and Low Carbon Transitions*. London: Routledge, 106–125.

Coutard, O., and Rutherford, J. (2016). Beyond the networked city: An introduction. In Coutard, O., and Rutherford, J. (eds.), *Infrastructure Reconfigurations and Urban Change in the North and South*. London: Routledge, 1–25.

Evans, J., Jones, R., Karvonen, A., Millard, L., and Wendler, J. (2015). Living labs and co-production: University campuses as platforms for sustainability science. *Current Opinion in Environmental Sustainability*, 16, 1–6.

Evans, J., and Karvonen, A. (2011). Living laboratories for sustainability: Exploring the politics and epistemology of urban transition. In Bulkeley, H., Castán Broto, V., Hodson, M., and Marvin, S. (eds.), *Cities and Low Carbon Transitions*. London: Routledge, 126–141.

Evans, J., and Karvonen, A. (2014). 'Give me a laboratory and I will lower your carbon footprint!' – Urban laboratories and the pursuit of low carbon futures. *International Journal of Urban and Regional Research*, 38, 413–430.

Evans, J., Karvonen, A., and Raven, R. (2016). The experimental city: New modes and prospects of urban transformation. In Evans, J., Karvonen, A., and Raven, R. (eds.), *The Experimental City*. London: Routledge, 1–12.

Evans, J. P. (2011). Resilience, ecology and adaptation in the experimental city. *Transactions of the Institute of British Geographers*, 36, 223–237.

Farías, I. (2011). The politics of urban assemblages. *City*, 15, 365–374.

Feola, G., and Nunes, R. (2014). Success and failure of grassroots innovations for addressing climate change: The case of the Transition Movement. *Global Environmental Change*, 24, 232–250.

Gibbons, M., Limoges, C., Nowotny, H., Schwartzman, S., Scott, P., and Trow, M. (1994). *The New Production of Knowledge: The Dynamics of Science and Research in Contemporary Societies*. Thousand Oaks, CA: Sage.

Gleeson, B. (2012). 'Make No Little Plans': Anatomy of planning ambition and prospect. *Geographical Research*, 50, 242–255.

Gopakumar, G. (2014). Experiments and counter-experiments in the urban laboratory of water-supply partnerships in India. *International Journal of Urban and Regional Research*, 38, 393–412.

Graham, S., and Marvin, S. (2001). *Splintering Urbanism: Networked Infrastructures, Technological Mobilities, and the Urban Condition*. London: Routledge.

Gross, M. (2010a). The public proceduralization of contingency: Bruno Latour and the formation of collective experiments. *Social Epistemology*, 24, 63–74.

Gross, M. (2010b). *Ignorance and Surprise: Science, Society, and Ecological Design*. London: MIT Press.

Halpern, O., LeCavalier, J., Calvillo, N., and Pietsch, W. (2013). Test-bed urbanism. *Public Culture*, 25, 271–306.

Harris, E. (2015). Navigating pop-up geographies: Urban space–times of flexibility, interstitially and immersion. *Geography Compass*, 9, 592–603.

Hoffmann, M. J. (2011). *Climate Governance at the Crossroads: Experimenting with a Global Response after Kyoto*. Oxford: Oxford University Press.

Karvonen, A. (2016). Low-carbon devices and desires in community housing retrofit. In Bulkeley, H., Stripple, J., and Patterson, M. (eds.), *Towards a Cultural Politics of Climate Change: Devices, Desires and Dissent*. Cambridge: Cambridge University Press, 51–65.

Karvonen, A., Evans, J., and van Heur, B. (2014). The politics of urban experiments: Radical change or business as usual? In Hodson, M., and Marvin, S. (eds.), *After Sustainable Cities?* London: Routledge, 104–115.

Karvonen, A., and van Heur, B. (2014). Urban laboratories: Experiments in reworking cities. *International Journal of Urban and Regional Research*, 38, 379–392.

Kshetri, N., Alcantara, L. L., and Park, Y. (2014). Development of a smart city and its adoption and acceptance: The case of New Songdo. *Communications & Strategies*, 96, 113–145.

Kullman, K. (2013). Geographies of experiment/experimental geographies: A rough guide. *Geography Compass*, 7, 879–894.

Laurent, B. (2011). Technologies of democracy: Experiments and demonstrations. *Science and Engineering Ethics*, 17, 649–666.

Lindsay, G. (1 February 2010). Cisco's big bet on New Songdo: Creating cities from scratch. *Fast Company Magazine*. Accessed 6 February 2017, www.fastcompany.com/1514547/ciscos-big-bet- new-songdo-creating-cities-scratch

Marvin, S., and Silver, J. (2016). The urban laboratory and emerging sites of urban experimentation. In Evans, J., Karvonen, A., and Raven, R. (eds.), *The Experimental City*. London: Routledge, 47–60.

McFarlane, C. (2011a). The city as a machine for learning. *Transactions of the Institute of British Geographers*, 36, 360–376.

McFarlane, C. (2011b). The city as assemblage: Dwelling and urban space. *Environment and Planning D: Society and Space*, 29, 649–671.

Meadowcroft, J. (1999). Planning for sustainable development: What can be learned from the critics? In Kenny, M., and Meadowcroft, J. (eds.), *Planning for Sustainability*. London: Routledge, 12–38.

Meadowcroft, J. (2009). What about the politics? Sustainable development, transition management, and long term energy transitions. *Policy Sciences*, 42, 323–340.

Middlemiss, L., and Parrish, B. D. (2010). Building capacity for low-carbon communities: The role of grassroots initiatives. *Energy Policy*, 38, 7559–7566.

Miles, M. (2007). *Urban Utopias: The Built and Social Architectures of Alternative Settlements*. London: Routledge.

Neal, P. (ed.) (2003). *Urban Villages and the Making of Communities*. London: Taylor & Francis.

Neilson, B., and Rossiter, N. (2008). Precarity as a political concept, or, Fordism as exception. *Theory, Culture & Society*, 25, 51–72.

Németh, J., and Langhorst, J. (2014). Rethinking urban transformation: Temporary uses for vacant land. *Cities*, 40, 143–150.

Nevens, F., Frantzeskaki, N., Gorissen, L., and Loorbach, D. (2013). Urban transition labs: Co-creating transformative action for sustainable cities. *Journal of Cleaner Production*, 50, 111–122.

Nowotny, H., Scott, P., and Gibbons, M. (2001). *Re-thinking Science: Knowledge and the Public in an Age of Uncertainty*. Malden, MA: Polity.

Okereke, C., Bulkeley, H., and Schroeder, H. (2009). Conceptualizing climate governance beyond the international regime. *Global Environmental Politics*, 9, 58–78.

Pickerill, J. (2016). *Eco-Homes: People, Place and Politics*. London: Zed Books.

Pickerill, J., and Maxey, L. (2009). Geographies of sustainability: Low impact developments and radical spaces of innovation. *Geography Compass*, 3, 1515–1539.

Ramadier, T. (2004). Transdisciplinarity and its challenges: The case of urban studies. *Futures*, 36(4), 423–439.

Reno, J. (2011). Managing the experience of evidence: England's experimental waste technologies and their immodest witnesses. *Science, Technology & Human Values*, 36, 842–863.

Roodhouse, S. (2010). *Cultural Quarters: Principles and Practice*. Bristol: Intellect Books.

Sabel, C. F., and Zeitlin, J. (2008). Learning from difference: The new architecture of experimentalist governance in the EU. *European Law Journal*, 14, 271–327.

Sabel, C. F., and Zeitlin, J. (eds.) (2010). *Experimentalist Governance in the European Union: Towards a New Architecture*. Oxford: Oxford University Press.

Sanderson, I. (2002). Evaluation, policy learning and evidence-based policy making. *Public Administration*, 80, 1–22.

Shapin, S., and Schaffer, S. (1985). *Leviathan and the Air-Pump: Hobbes, Boyle, and the Experimental Life*. Princeton, NJ: Princeton University Press.

Silver, J. (2014). Incremental infrastructures: Material improvisation and social collaboration across post-colonial Accra. *Urban Geography*, 35, 788–804.

Simone, A. M. (2013). Cities of uncertainty: Jakarta, the urban majority, and inventive political technologies. *Theory, Culture & Society*, 30, 243–263.

Späth, P., and Rohracher, H. (2012). Local demonstrations for global transitions: Dynamics across governance levels fostering socio-technical regime change towards sustainability. *European Planning Studies*, 20, 461–479.

Tonkiss, F. (2013). Austerity urbanism and the makeshift city. *City*, 17, 312–324.

Vanolo, A. (2013). Alternative capitalism and creative economy: The case of Christiania. *International Journal of Urban and Regional Research*, 37, 1785–1798.

Voytenko, Y, McCormick, K., Evans, J., and Schliwa, G. (2016). Urban living labs for sustainability and low carbon cities in Europe: Towards a research agenda. *Journal of Cleaner Production*, 123, 45–54.

Walker, G. (2011). The role for 'community' in carbon governance. *Wiley Interdisciplinary Reviews: Climate Change*, 2, 777–782.

12

Experiments and Beyond

An Emerging Agenda for Climate Governance Innovation

BRUNO TURNHEIM, PAULA KIVIMAA AND FRANS BERKHOUT

12.1 Introduction

This volume has discussed the role of experimentation in climate governance. Recognising the growing significance of experimentation in political and socio-technical responses to climate change, we have been interested in understanding how experiments in climate governance can lead to broader changes in the rules, practices, norms and arrangements constituting responses to climate change in widely differing settings. We have sought to move the analysis of climate governance experiments beyond a focus on single cases and to develop an argument for seeing experimentation as a new habitus in climate governance, defining the character of a range of responses which evoke new puzzles for climate governance. If experimenting is a new means of conceiving of and doing climate governance, then we also expect it to have broader outcomes beyond the moment of experimentation. Experiments are a means to an end, even if they are a continuous feature of political responses to climate change. Such an experimental conception of climate governance poses a set of major conceptual challenges. And this volume has sought to clarify what these are and to make contributions to their understanding and resolution.

Convinced that a creative dialogue between governance and policy studies and studies of socio-technical innovation would help in reframing experimental climate governance, we have sought in this volume to provide a forum for such a dialogue. Both fields have had complementary but largely disconnected debates about the generation of new ideas, the role of entrepreneurial activity and the wider adoption and diffusion of new ways of doing things. By encouraging a connection and a mutual grappling with each other's concepts and case studies, we believe that we have made some progress in developing a basis for richer understandings in both fields.

We have been concerned with a number of specific questions:

• What do climate governance experiments lead to beyond their particular contexts, and do they influence changes in norms, incentives, rules, behaviours, relationships and arrangements for addressing climate change mitigation and adaptation?
• If experiments are largely uncoordinated and entrepreneurial initiatives by new coalitions of actors, what direct outputs do the experiments produce (for example, ideas, networks, capabilities and narratives), and how do they come to have broader influence?
• What notions of diffusion, reproduction and embedding can best describe the process by which the multiple possible outputs of experiments come to generate broader outcomes?

In trying to answer these questions, we have been particularly interested in what we see as a significant gap between experimental activities and outputs, on the one hand, and wider outcomes and more lasting impacts of experimentation on the other. We have problematised this as a process of embedding the outputs of experiments in wider policy systems, and we have sought to explore various forms and underlying mechanisms for this process of embedding.

In this concluding chapter, we return to the challenges and questions posed in our introduction. First, we noted a fragmented landscape of activities designated as experiments coupled with greater interest in experiments in climate governance as well as a growing ambiguity about what constitutes experiments and about their distinctive features as governance mechanisms. Therefore, in the first part of this chapter we aim to clarify what we believe constitutes a governance experiment before assessing the process by which experiments come to have a broader influence through embedding. In this latter analysis, we are especially interested in actors and agency and how this shapes and influences what happens beyond singular experiments. Second, we argued in the introduction that given the growing interest in experimentation, little attention has been paid to how experiments can contribute to generating wider influence and change towards low-carbon and climate-resilient societies at the level of broader political and economic systems. This has been a core focus of this volume. Thus, in the second part of the conclusions, we highlight what it is that gets embedded and illustrate four different perspectives on the processes by which embedding can take place.

Third, we consider climate governance as a whole, including public, private and civic elements of governance, and assess what experiments may mean for societal responses to climate change in the round. In particular, how do experiments fit into climate governance arrangements? Can they generate significant policy changes and outcomes that contribute to mitigating and adapting to climate change? Can we identify factors that lead to successful embedding of the learning and resources afforded by experiments? The concern here is about whether climate governance experiments will have more lasting impacts on climate governance, leading to the decarbonisation of societies and building resilience, especially for more vulnerable communities. It may be early to have clear evidence about such effects, but if we are to anticipate them, we need some conceptual apparatus for tracing through such impacts.

12.2 Clarifying What We Mean by Experiments and Embedding

Many of the claims to experimentation within climate governance hold the potential for bringing about radical new ways of dealing with deep structural problems linking climate change to how our economies and societies are organised. In practice, there is considerable variety in the kinds of initiatives that can be considered as climate governance experiments (see Chapter 1). This is both a valuable source of richness and a source of confusion that creates a risk of distraction away from some of the core problems and objectives of climate governance.

Before we can understand fully the transformative potential of climate governance experiments, we need to consider some crucial aspects of experimentation, inherent differences in the various approaches to experimentation that are being put forward and some of the critical challenges they face. Further specifying the different ways in which climate governance experimentation can be configured will enable us to understand better the relevance and means of their embedding.

Regardless of how defined and circumscribed any given experiment is, the focus of this volume has compelled us to think about boundaries and their crossing: when does an experiment begin and particularly when does it end? This question of boundaries becomes even more salient as it informs the kinds of obstacles, challenges and opportunities at play *beyond* experiments. In practice, experimentation in governing is not always neatly delimited by a boundary that simply needs to be overcome. There is a degree of 'messiness' involved in experimental practices and processes of their embedding, and this needs to be a topic for investigation if we are to say something useful about climate governance experiments and their multiplication, scaling, diffusion or reproduction. The likely explosion of 'climate experiments' will only make this sense-making challenge more salient.

When addressing this core problem, we have asked contributors to be explicit about what an experiment is, what it is intended for and when it becomes something other than an experiment. This means attending to the question of boundaries when defining governance experimentation and being explicit about the difference between an experiment and the context of its embedding, as well as the object of such embedding – whether this is the specific experiments themselves or their outputs (see Section 12.2.1). This has led some authors to focus on the articulation of the boundary between experiments and their wider embedding, mobilising, for instance, hinge notions such as the 'pilot paradox' (van Buuren et al., Chapter 8), the 'twin notions' of 'anchoring and mobility' (Carvalho and Lazzerini, Chapter 3) or the seemingly paradoxical notion of 'permanent experiment' (Karvonen, Chapter 11). So, empirical studies recognise the inherent differences and resulting tensions between experiments and their embedding. There is a time, place and mode for experimentation, which structure how experiments are done, and how they can be evaluated. Equally, there is a time, place and mode for the embedding of experiments beyond the experimental context. These distinctions may differ according to the conceptual perspectives mobilised and the nature of experiments being considered. Nevertheless, they point to the need to problematise the boundaries between spaces of experimentation and spaces of embedding, as they may be in considerable tension.

In this section, we seek to bring further clarity as to the boundaries of experiments (when/where do they begin, and when/where do they end), the variety of actors involved and the claimed motives for climate governance experimentation. We consider the relevance of these boundaries, taking account of evident ambiguities, before developing an understanding of how such boundaries come into tension as experiments become more than experiments. As much as is possible, we try to clarify how boundaries come to be articulated, and how they are redefined and transgressed as the outcomes of experiments become more widely embedded.

12.2.1 Constructing Experiments

In the introductory chapter, we offered a number of metaphors to make sense of the multiplicity of ways in which experimentation is conceived in governance and policy studies and innovation studies. We also posited that experiments tend to be subject to pre-determined temporal, spatial and evaluative boundaries. Indeed, most formal experiments have (a) a beginning and an end tied to a budget, activities and expected outputs, (b) a particular institutional and geographical context within which experimental activities are carried out and (c) clear goals and means for their evaluation. These boundaries are seen as essential to defining the exceptional and distinct character of formal experiments – as close as possible to the metaphor of the lab, from which generalisable knowledge can be extracted. This suggests that as an experiment crosses these boundaries, it becomes more than an experiment (see Figure 12.1).

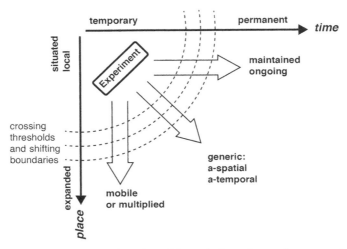

Figure 12.1. Temporal and place-based boundaries of experiments and paths to their crossing.

Experiments range from those specifically initiated as experiments with the intention of *experimenting* (what can be regarded as formal experiments), to more ad hoc and informal kinds of experimental climate action (i.e. informal experiments). Regardless of where experiments are positioned along this spectrum, Hoffmann (2011: 57) argues that experiments should not be considered as 'simply random and idiosyncratic initiatives'. In each case, we are considering discrete initiatives carried out by individuals, organisations or collectives that seek to address specific climate governance problems in ways that are not fully captured by conventional responses and arrangements.

At one end of the spectrum, ad hoc experiments may be relatively spontaneous and isolated (Castán Broto and Bulkeley, Chapter 4; Hölscher, Frantzeskaki and Loorbach, Chapter 7), emerging in multiple places as the result of favourable conditions (Karvonen, Chapter 11). Such experiments may provide common directions in practice communities or niches. They may not be labelled as experiments by their proponents, tending to have an emergent and situated character, whether or not they have a collective purpose. They stretch conceptions of experimentation as being inherently deliberate (cf. Kivimaa et al., 2017). The start and end points of such experiments may be difficult to distinguish. Further, they are more likely to produce emergent solutions to problems – by diverging from existing practices and routines – than being organised to specific ends, such as a more climate-resilient city or a region.

At the other end are *deliberate*, *formalised* and *structured* experimental settings, often initiated by established governance actors (e.g. public authorities,

municipalities, industry, non-governmental organisations [NGOs]) with the objective to learn from a variety of 'pilots' and demonstrations (van Buuren, Chapter 8;
Nair and Howlett, Chapter 9; Heiskanen and Matschoss, Chapter 10; Karvonen,
Chapter 11). For these more structured settings – still quite rare in climate
governance – fragmentation may be less of a problem since experimentation is
highly directed and inscribed in a deliberate strategy. Such experiments tend to be
more defined and bounded. They can be problem- or change-oriented, and desired
end states are easier to recognise. Yet, evaluations of 'success' have so far tended
to focus more on the level of outputs than the substantial broader outcomes they
may have generated (Pallett, Chapter 5; van Buuren et al., Chapter 8; Heiskanen
and Matchoss, Chapter 10). Table 12.1 illustrates a nomenclature of experiments as
initiated by public or other actors and according to the extent of formal structure.
The different dimensions considered need to be seen as extremes of formal and
informal experiments and experiments initiated by different types of actors.

If the boundary of experiments can prove elusive, then the boundary between
experiments and their embedding may be even more difficult to pin down. However, if we start from an assumption that the experimental does give way at some
point to the post-experimental, then we also assume that there are temporal, spatial
and socio-cognitive markers that indicate that an experiment is being embedded
beyond its initial configuration. In the Introduction we argued that this would be (a)

Table 12.1. *Mapping the range of climate governance experiments*

		Initiating/lead actors	
		State (local, national or regional governments)	Non-state (citizens, NGOs, businesses)
Degree of formalism	Formal and defined	Policy instrument pilots Policy process experiments Public engagement experiments	Transition management and arenas Bounded socio-technical experiments Novel partnering networks acting as innovation platforms Living labs
	Informal open ended	Experimental governing frameworks based on more recursive rule setting and implementation Policy innovations not clearly defined as experiments and limited in terms of time and scale	New governance concepts Grassroots innovations Experimental business models Climate projects and initiatives Sustainability experiments

222 Bruno Turnheim, Paula Kivimaa and Frans Berkhout

when an experiment is continued long enough or beyond an initially set end date; (b) when an experiment becomes larger, is multiplied or is replicated beyond the scope of its initial form or (c) when the knowledge and learning that an experiment generates is adopted in the actions and routines of people beyond the experiment, or institutionalised in ways of governing beyond the experiment.

12.2.2 Actors and Agency in Experiments

The range of actors involved in experiments extends from public policy actors (political decision-makers, civil servants and public agencies), through research organisations to a variety of private actors (see, for example, Pallett, Chapter 5; Farrelly and Bos, Chapter 6; Heiskanen and Matchoss, Chapter 10). Throughout the 'unpredictable career of climate experiments', external actors crop up to become involved in experimentation, whether out of genuine interest, aspirations for reaping direct benefits or associated reputational gains of 'flagship' initiatives. This may fulfil an effective function in terms of resource mobilisation by enhancing visibility and further articulation of successful initiatives. Yet, it is also a symptom of asymmetries in risk-taking between the initiators of experimentation and potential beneficiaries reaping the rewards of experimentation, including the potential for co-option and predation. The more visible climate experiments tend to attract more external interest and resources, irrespective of their actual climate benefits, as they may serve other political and commercial interests as well. The co-optation of experiments by others can be critical to ensuring the viability of initiatives through time and beyond initial funding periods, but may also lead to disengagement by initial members, as it 'stretches' their initial framing.

Contributions to this volume offer a variety of insights on who and what is seen as mediating and intermediating the embedding of experiments, and mobilising different types of boundary-spanning processes. Actors are central to embedding experiments but these actors may differ from those who initiate and undertake experiments. Heiskanen and Matschoss (Chapter 10) argue that, within experiments, its actors may 'have diverse or even divergent understandings of what is being tested'. Several authors differentiate between insiders and outsiders to experiments. Insiders comprise the 'coalition of the willing' that not only have the power to form and carry out experiments but also need sufficient leverage to influence the embedding of the outputs of experiments (Hölscher et al., Chapter 7). Outsiders would not have a central role in the experiment itself, but may prove highly important for its embedding. Such actors, including innovation, niche or transition intermediaries (Kivimaa, 2014) can aggregate learning and other outputs of experiments on a higher level, communicate such learning and outputs to stimulate the replication of experiments or influence the emergence of new supportive institutional structures (Geels and Deuten, 2006; Geels and Raven, 2006;). Carvalho and Lazzerini (Chapter 3) show how Community Chance Aggregation concepts and practices became progressively legitimated and transformed in California by multiple actors (consultants, not-for-profit organisations) who become intermediaries 'actively and purposefully moving the concept across space and initiating new rounds of mobility'.

12.2.3 Motives and Objectives of Actors

In Chapter 1, we provided an overview of motivations for experimentation by exploring some common metaphors: testing hypotheses, setting designs, learning by doing, creating radical novelty, nurturing and ordering collective reality. These narrative frames link to the specific objectives of experimentation in different settings, the definition of 'success' and 'failure' and how these may be evaluated. Contributions to this volume have considered broader expectations and outcomes linked to experimentation, going beyond these metaphors, and how they come to be realised and produced.

Because most climate governance experiments are framed as projects, initiatives or pilots, they tend to involve rather practical definitions of objectives, targets and activities. One of the features of an experiment is that it must come to an end; perpetuation and transfer of outcomes is inherent to the motives of actors involved in experiments. Such practical considerations can be highly informative as to the starting motives and purposes of experiments. But it is in the nature of experiments that they mutate and change as they unfold. Objectives, targets and activities may change over time, as initiatives come to be reframed in the face of real-world challenges (see, for instance, Pallett, Chapter 5). Given the preliminary, provisional, untested and entrepreneurial nature of experimental configurations, and since the primary objective is to learn by doing and to try again, mobility and adaptation may be inherent to social experimental settings. In practical terms, the resources and motivation available to experimental actors are likely to be highly dependent on external sponsors and factors. Perpetuation and transfer of outcomes will also depend on these sponsors, making experimental actors sensitive to their interests and perspectives and modifying objectives, narratives and promises in tune with modifications in such interests and perspectives. Departures from the original premises of an experiment may already point to the start of a process of embedding lessons, resources and networks. Most views on experimentation emphasise learning as a key motivation. Such views range from simple 'testing' to internalising new ways of thinking, doing, acting and knowing. The latter may be a key step towards more transformative change.

In our original definition, experiments are always intended to have outcomes that can be transferred and adopted elsewhere. But there is a range of conceptions of what these outcomes may be, and different actors may hold differing and inconsistent views about what these outcomes will be. The simplest view is that an experiment will demonstrate a way of doing things so that it can be adopted more or less unaltered by others. If experiments are demonstrations, then they represent microcosms that are complete, fully ordered and functional and mobile so that they can be transposed to other contexts. The problem of replication will involve developing the needed resources, following the rules and learning how to do things, including education and training.

But the outcomes of experiments and their transfer is often more complex and contentious. Several contributions to this volume (Carvalho and Lazzerini, Chapter 3; Farrelly and Bos, Chapter 6; Hölscher et al., Chapter 7; van Buuren et al., Chapter 8) suggest a further dimension, that of the degree of *structuration* involving the abstraction of lessons, generalisation of knowledge and the modification of these lessons to new contexts. Experiments in this case are less microcosms to be emulated so much as examples which become generative analogies for others, clarifying problems, outlining general principles and providing encouragement that alternatives may be viable. Such analogies may also suggest different pathways for embedding, involving a shift from situated and particular action/knowledge to expanded relevance (diffusion) and/or structuration (abstraction, generalisation and institutionalisation). Most processes of embedding may include a combination of emulation and abstraction. For instance, Strategic Niche Management talks of the need for experiments to acquire generic and mobile features so that knowledge can be aggregated and circulated amongst relevant communities (Geels and Deuten, 2006). The embedding literature talks of embedding, dis-embedding and re-embedding (Geels and Deuten, 2006), while contributions in this volume discuss processes of anchoring, circulation and re-anchoring (Carvalho and Lazzerini, Chapter 3; Castán Broto and Bulkeley, Chapter 4). Expanding on the idea of the diffusion of experiments through time and organisational space (see Figure 12.1), we here add a third dimension of structuration (see Figure 12.2). We believe embedding should be viewed as being more than the emulation, diffusion or scaling of exemplars demonstrated in experiments; it should also include processes of abstract lesson-learning, adaptation and reconfiguration of governance systems.

We see embedding as typically including some more or less radical process of 'fitting' experiments and their outputs to post-experimental contexts. In this process, experiments tend to be reframed, co-opted and altered to become more adapted to the existing ways of doing things and of norms and structures. As they do so, their transformative influence may become more limited (Smith and Raven, 2012; van Buuren et al., Chapter 8), both in the sense of being modified to fit the

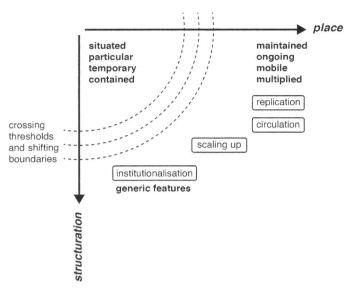

Figure 12.2. Experiment embedding as moving through spatial and structuration levels.

new host governance regime and in the sense that additional entrepreneurial and innovative work has been done, leaving the imprint of the experiment to be fainter. For example, Nair and Howlett (Chapter 9) and Carvalho and Lazzerini (Chapter 3) analyse the 'journeys' of specific policy experiments over time and place, through examples of community electricity procurement, insurance schemes and development projects for farmers in relation to climate change adaptation. This view places emphasis on embedding the outcomes of experimentation (e.g. Castán Broto and Bulkeley, Chapter 4; Farrelly and Bos, Chapter 6), including visions, narratives and insights about how change works, new governance rules and new businesses or citizens practice (Kivimaa et al., 2017). This mixed pattern of outcomes, a combination of the 'demonstration' and the 'analogy' views of potential outcomes, is recognised by actors involved in experiments and their embedding. This was explored in chapters that looked at the interface between isolated experimental outputs and their more aggregate outcomes (Hölscher et al., Chapter 7; van Buuren et al., Chapter 8; Carvalho and Lazzerini, Chapter 3).

In the context of sustainability transition studies and niche protection and development, Smith and Raven (2012) propose a useful distinction between strategies of 'fit and conform' as opposed to 'stretch and transform', indicating differing degrees to which niches and the experiments they contain fit to or transform existing selection environments. Applied to climate governance experiments, 'fit and conform' empowerment narratives suggest that embedding is not transformational, as experimental outcomes adapt to existing governance and socio-technical structures. Conversely, the narratives of 'stretch and transform' describe a larger extent of change in which the practices, technologies, visions and rules proposed by the experiments lead to a reconfiguration of existing practices and structures.

12.3 Embedding Experiments: Paths, Processes and Challenges

12.3.1 From Outputs to Outcomes of Experiments

Whether part of a highly formalised and structured experimental *dispositif* or a more ad hoc and emergent process of adaptation and reconfiguration, climate governance experiments can be qualified by the outputs and outcomes they produce. The embedding of experiments raises a number of issues for evaluation: given the range of outcomes and pathways, how should objectives be stated and what should be the success criteria. There is a need for clarifying what experiments 'produce' as their outputs and outcomes, individually and in the aggregate.

The outputs of experiments are often described operationally in terms of specific activities and the fulfilment of objectives. The expected outputs range from market impact data (Tassey, 2014) to new, shared visions about future socio-technical systems (Kemp, Rip and Schot, 2001). 'Knowledge about how something (e.g. a technology, a service, a policy, etc.) "works in the real world"' (Kivimaa et al., 2017: 2), is often central to an experiment's justification (cf. Kemp et al., 2001; Brown and Vergragt, 2008; Tassey, 2014). In formal public governance, new goals, new instruments or new types of leverage mechanisms and implementing organisations could be expected as outputs of governance experiments (Jordan and Huitema, 2014; Upham, Kivimaa, Mickwitz et al., 2014). In informal governance settings, outputs of experiments may be less well-defined and associated with alternative modes of governing to more formal and established ways (Hoffmann, 2011). Kivimaa et al. (2017), on the basis of a systematic review of transition experiments literature, identify six kinds of outputs: discursive (new visions or integration of perspectives), technological (practical applications of new technologies), infrastructural (changes in planning for land use and infrastructure build-up), policy related and institutional (new planning practices, new actors in policymaking and changes in responsibilities in multi-level governance), business related (new business models) and citizen oriented (improved citizen engagement and emergence of alternative communities). Novelty also springs from more open search processes, such as in the case of 'failed experiments' that do not produce the expected outputs, but may reorient the search process in fruitful directions, by excluding certain paths or allowing focus on others. The history of innovation may be seen as a history of unintended discoveries and surprises stimulated by unsuccessful experimentation (Abernathy and Utterback, 1978). But as we have seen, whether this is because of the failure to achieve an original promise or a failure to adapt to changing interests of sponsors and mutating selection environments is often hard to tell.

It is often the outputs – rather than the broader impacts and outcomes the experiments generated – that are used in measuring the success or failure of experiments. The absence of a consideration of outcomes in assessing experiments, which are important to embedding experiments, may be explained by the necessary investment of time and effort. Outcomes may be *intended* and linked to a purposeful process of realising specific expectations in practice, but also *unintended* surprises, traces and co-produced transformations of the surrounding context, participants, routines and practices. They range from short-term to longer-term. Short-term outcomes may come in the form of new narratives, policies, business practices and networks, knowledge as to what factors best promote change or even replication of the experiment in another setting (Kivimaa et al., 2017). Long-term outcomes, such as broader socio-technical change in markets or practices, may be initiated or contributed to by experiments (Brown and Vergragt, 2008; Berkhout et al., 2010).

In broad terms, we would bring attention to three kinds of outcomes of experimentation:

1. Practical or abstract knowledge and learning
2. Systemic and transformative change
3. Political ordering and mobilisation

First, a major objective and co-produced outcome of experimentation is new knowledge and learning (see, for instance, Farrelly and Bos, Chapter 6; Heiskanen and Matschoss, Chapter 10). This may be related to more conventional views of experimentation as testing and learning by doing, where the object of learning concerns the validation of claims about the external world and the means by which we may exert control on it. It may also concern knowledge about uncertainties and the limits of control and predictability. Retaining knowledge and learning derived from specific experiments is crucial to supporting their transposition elsewhere, their translation into more generic and mobile forms (e.g. via 'recipes' and 'templates') or the doing of further experiments along a given path. Beyond being a valuable currency, learning and knowledge stem from *social processes*. Crucial to the transformative potential of experimentation is 'second-order' learning (Brown and Vergragt, 2008), or the ability to learn how to learn. Yet, obtaining evidence of this as an outcome of experiments is difficult (Kivimaa et al., 2017). The fact that learning is always enacted by particular actors underlines the importance of experimentation as a capability: 'hosting' experiments becomes crucial to building up expertise and innovative capabilities, as well as the capacity and routines to learn from multiple experiments.

A second outcome of experimentation is the capacity to transform the governance milieu through which it flows (see, for instance, Castán Broto and Bulkeley, Chapter 4; Pallett, Chapter 5). By transformation, we mean more discontinuous or radical changes in institutional structures, actor behaviours and problem frames. Beyond local transformation, this may refer to a multiplicity of processes that together contribute to radically new configurations or the reorientation of paths. Hoffmann (2011: 29) describes two ways in which climate governance experiments can disrupt the status quo to influence systemic transformative change: acting as 'first, a source of friction in politics and markets that catalyses demand for broad transformation in societies and economies, and second, a source of smoothing that provides the technologies and institutions to respond to the demand for transformation'. Pursuing this kind of systemic transformation objective implies quite different criteria by which to evaluate the outcomes of experiments, as it requires a shift from a focus on 'learning something new' to 'enabling and sustaining systemic change' when setting priorities and designing experiments. This may be an aspirational motivation for experiments, even if weakly signalled by initiators and sponsors, and this may explain obstacles to embedding, as suggested by van Buuren et al. (Chapter 8). That said, the transformative impact of experiments may also be overstated. While policy experiments can break institutional rigidities, to stimulate long-term governance changes, they need to be supported by radical changes in formal regulations and professional standards (Primmer et al., 2013). The potential for transformation may trigger experimentation, which in turn is sustained by other aligned changes to constitute transformation jointly.

Third, there may be broader political outcomes of governance experimentation through the reconfiguration of collective political and institutional orders. Any one of the metaphors on experimentation considered in Chapter 1 is the carrier for a certain appreciation of the world, our means of control over it and the principles and technologies by which it may be governed. Climate governance experiments produce and reproduce tools, associated norms, values and meanings and sociopolitical orders – which can promote the status quo or fundamentally rearrange collective realities. Experiments may be powerful tools for the empowerment of individuals and collectives as they open up new spaces for intervention beyond and across usual jurisdictions (Hoffmann, 2011), provide opportunities for the accumulation of expertise around new ways of handling climate governance and the creation and assertion of new roles (e.g. progressive advocates of change, intermediaries, system builders and involved users). Experimentation is also used for more instrumental and potentially contested purposes. Demonstrations and pilots often focus on producing highly visible exemplars that serve additional purposes, such as enrolling external support, producing a positive image of the actors involved (as doing something about problems or being at the vanguard of developments) or contributing to hopeful narratives and boosting confidence in the future viability of a specific option (see, e.g., Hölscher et al., Chapter 7). Nair and Howlett (Chapter 9) also note that politics become more important determinants of the success of 'policy pilots' than the design of such pilots. Such outcomes may explain the apparently disproportionate costs of some flagship projects.

Expectations about the lasting outcomes of experimentation have consequences for how embedding may happen or how it can be evaluated. Because of our interest in embedding, we focus beyond the more immediate and project-specific success criteria often deployed to evaluate the outputs of discrete experiments. Instead, we suggest moving towards evaluations that enable comparison across different experimental contexts and which are designed to assess progress towards more systemic outcomes and processes. This leads us to question *what* it is that becomes embedded, and *how* this can be evaluated.

12.3.2 *Processes of Embedding*

Experimental activities are often understood as located upstream of longer sequences of change-oriented processes. For example, Grin, Rotmans and Schot (2010), in connection to transition studies and transition experiments, talk about three different processes that can follow from experimentation: *deepening* refers to learning as much as possible from the transition experiment in question, *broadening* refers to repeating an experiment in an adjusted form in a different context and *scaling up* to embedding an experiment in the existing structures, practices and narratives of an incumbent regime. Expressed as an outcome, deepening may be viewed as shifts in ways of thinking and of practices, as well as the organisation of physical, economic and institutional structures. Broadening is the replication of the experiment in different contexts, whereas scaling would influence established ways of thinking, doing and organising (Grin et al., 2010; Raven, Van Den Bosch and Weterings, 2010; Bos and Brown, 2012).

While this categorisation is a useful starting point, from the contributions of this volume, we see an emergence of new notions of embedding, including 'circulation' (Castán Broto and Bulkeley, Chapter 4) and a more critical view of replication and scaling up (e.g. Carvalho and Lazzerini, Chapter 3). Drawing on the insights from the chapters in this volume, we suggest that embedding can be captured under four macro-processes: scaling up, replication, circulation and institutionalisation, summarised in Table 12.2 and Figure 12.2. We see learning

Table 12.2. *Processes and elements of embedding experiments*

Object of embedding (what becomes embedded?)	Embedding processes			
	Scaling up	Replication	Circulation	Institutionalisation
Experiment as a blueprint	Increase of the scope and length of the experiment (possibly ending the experiment status)	Creation of a new context and location-specific application of the experiment	No blueprint exists	Experiment continues and stops having an experimental status.
Experiment as a source of knowledge and learning	Knowledge and learning becomes mainstreamed in the process of scaling	Diffusion and (re-)contextualisation of codified and practical knowledge	Flow, reuse and modification of knowledge and learning	Institutionalisation of knowledge and learning is generated by the experiment in form of rules, practices and scripts.
Experiment as a source of new policies	Experiment's policy-relevant output/outcome taken to more permanent and broader scale use	Replication, with adaptation, of an experiment's policy output/outcome to new contexts	Experimentation as a governance approach	Experiment's policy output/outcome becomes embedded in formal and informal governance structures.
Experiment as network development	Expansion of the (whole) actor network initiated by the experiment	Replication of an experiment's actor configuration, modified to new local circumstances	Actor networks evolve as an outcome of the flow of ideas and resources generated by experiments.	The experimental actor network configuration replacing formal and informal governance structures
Experiment as a source of new socio-technical configurations	Broader uptake of the technological (or service) outputs of experiment	Diffusion and (re-)contextualisation of the technological (or service) configuration	Socio-technical configurations as a contingent outcome of circulating governance narratives and norms.	Technological (or service) configuration becomes a widely accepted solution.

Experiment as a stimulus for new practices	Mainstreaming of new practices emerging or proven in the experiment	Replication of adapted practices associated with the experimental design *or* anchoring to local context through situated practices enabling the uptake of technological (or service) configuration.	Practices pioneered in experiment become part of an assemblage of circulating governance practices.	Practices initiated by the experiment become an established part of formal and informal governance structures.
Experiment as political reordering	New governance regime modelled in experiment replaces an existing regime.	Governance regime is modified through the emergence of experiment-based rules and routines in different places	Governance regime absorbs, processes and internalises experimental outcomes – impacts on governance order unclear	Formal and informal governance is adjusted systematically, including redistribution of administrative and legal power.
Experiment as a trigger to transformation	Experiment overturns an incumbent governance and associated socio-technical regime.	Experiment leading to chains of similar experiments that cumulatively change the socio-technical setting	Experimental approach to governance changing the system from within.	Experiment initiates a process of de-institutionalisation and re-institutionalisation.

as an integral part of all these processes with the process and nature of learning differing in each form of embedding.

In studies on socio-technical experiments, scaling up is rarely examined critically although there are differing interpretations of what it may imply (e.g. Jolly, Raven and Romijn, 2012). For example, Jolly et al. (2012) refer to scaling up as having a wider impact at the location of the original experiment, including achieving greater penetration of target groups and extending services to new people, among other things. Laakso, Berg and Annala, (2017), in their meta-study of twenty-five articles on experiments, mention scaling up as integrating and applying the experiment at a higher system level. Hartmann and Linn (2009) cover not only the expansion but also the replication and sustaining of experimental configurations over time in their definition of scaling up. The conceptualisation of scaling ranges from a narrow 'making the experiment bigger' to considering both scaling up within and across locations and sustaining experiments over time (see also Figure 12.1). What is less clear is what features of the experiment are being scaled and how scaling takes place from a pluralism of possible pathways.

Scaling up may be a process in which an experiment is (and its outputs are) expanded in scope (geographical, administrative, financial, etc.) or duration. This assumes that scaling always implies expansion. It also assumes that expansion – of technologies, practices, actor-configurations and rules – is associated with adoption into standard practice of these experimental outputs. Some of the contributions to this book talk about crossing scales and breakthrough thresholds (Hölscher et al., Chapter 7; Nair and Howlett, Chapter 9). For policy piloting (as in Nair and Howlett, Chapter 9) or technology piloting, scaling up can occur through the adoption of governance modes and instruments in larger jurisdictions. However, for some climate governance experiments – particularly those that happen in more local or polycentric settings – scaling up appears to be rare. The experimental scale may be the appropriate scale for a given intervention, and its broader impact will depend on processes of replication. More profoundly, scale is a relative rather than an absolute idea, implying that scaling always needs to be understood in relation to something else, like a prevailing governance regime.

Replication is often proposed as a contrast to scaling. It refers to repetition and reproduction of an experiment in a new context (e.g. another policy domain) or location (e.g. new city or country). Replicability is one of the main characteristics of conventional experiments. Replication in social and political settings where learning is a principle goal usually involves adaptation to new contexts, locations, interests and problem frames. Thus, replication carries with it an idea of change along the way. Carvalho and Lazzerini (Chapter 3) illustrate how in the process of replication an initial experiment changes and adapts to its new location. They talk about the individual journeys of specific solutions as '"spatialised" sequences of anchoring, recombination and mobility'. According to this view, the embedding of experiments is not a simple process of diffusion, expansion or replication of an exemplar, but a process of modification that negotiates an inherent tension between situated and mobile knowledge and practice. In effect, it requires experiments or their outcomes 'mobile' (Carvalho and Lazzerini, Chapter 3; Nair and Howlett, Chapter 9). Here, actors pursuing private or common goals, holding relational assets and engaging in deliberate interventions play an active role in overcoming the obstacles through active (re-)contextualisation and transformation of the governance intervention itself. While replication is often used to refer to the whole experiment, it may also entail specific tangible and intangible elements (such as a policy design or a technological configuration). Kivimaa et al. (2017) argue that this form of active and mobile replication is more typical of processes of embedding for climate governance than scaling up in the sense of expansion.

Circulation presents the embedding of experiments as the ongoing transformation and reconfiguration of existing regimes. According to Castán Broto and Bulkeley (Chapter 4), experiments can transform existing (urban) regimes by locally reconfiguring the circulation of socio-technical and socio-ecological flows. Fundamental to the lasting transformational impact of experiments on urban milieu is their ability to open new spaces for practicing a 'politics of hope'. The idea of circulation relates to the career of experiments seen as journeys (Carvalho and Lazzerini, Chapter 3; Pallett, Chapter 5; Heiskanen and Matchoss, Chapter 10). Castán Broto and Bulkeley (Chapter 4) suggest that circulation can be understood as involving (a) the movements of 'things' as calling for the realigning of existing relations (e.g. via new parameters and criteria), (b) a focus on those 'things' that can accompany or enable circulation of ideas or policies (e.g. templates, finance mechanisms and standards), and (c) amplification and dissemination processes that can extend the pool of receptive public (through e.g. the development of new imaginaries and visions that can support the mobilisation of expectations).

Institutionalisation can be described as a process in which an experiment, or its outputs or outcomes, becomes part of governance structures, rules, norms and routines, whether formal or informal. This can imply a variety of forms from standardisation of technologies (that have been developed through stages of experimentation) and professionalisation of new roles and activities (Pallett, Chapter 5), to the uptake and incorporation of new approaches into 'formal policies or implementation programmes' (van Buuren et al., Chapter 8). Institutionalisation can be a bottom-up and uncoordinated process, or be the outcome of a more directed process. In the latter, policy experiments can be understood as opportunities for the phased introduction of major government policies or programmes, 'allowing them to be tested, evaluated and adjusted before being rolled out nationally' (Cabinet Office, 2003; cited in Nair and Howlett, 2016: 69). Interestingly Nair and Howlett (Chapter 9) show that climate adaptation policy pilots in India have been integrated into existing mixes of policies mainly through incremental adjustments, while a process of institutionalisation as a form of embedding was only observed in one out of fourteen policy pilots. To succeed, institutionalisation may need to be preceded by a process where experiments deinstitutionalise or destabilise a prevailing governance regime before they can become embedded (cf. Turnheim and Geels, 2012; Kivimaa and Kern, 2016). Hölscher et al. (Chapter 7), by drawing on climate governance, sustainability transition and sustainability resilience literatures, depict two intertwined activities and related competences that are crucial for experimentation and its embedding: transforming and orchestrating. Their key argument is that transformative capacity held by those governing is necessary to achieve a high degree of novelty. Orchestrating capacity is crucial to create consensus about the direction of change and to mobilise action in a coordinated manner.

These different processes may coexist or they may occur in sequences. For example, replication may be a precursor of institutionalisation. The coexistence of plural processes better characterises the specificity and contextualisation of processes of embedding, as a social and political process. Overall, contributions to this volume show that when analysing and evaluating the embedding of experiments, we can distinguish between several features of experiments and a range of pathways along which governance experiments may act beyond their original project and time boundaries. These distinctions are organised in Table 12.2.

12.4 Implications for Climate Governance

The Paris Agreement (2015) emphasised the role of informal governance approaches and experimentation in tacking climate change. At present, a common issue with climate governance experiments is that, while typically problem-oriented, they are often isolated, fragmented and weak. To follow the path set by the Paris Agreement, there is thus a need to address the issue of fragmentation. It remains difficult to define where experiments may 'fit' within climate governance arrangements, how they may contribute to generate momentum for systemic changes in wider society and what is the scope for deliberate strategies to broaden their impact. We, therefore, suggest complementing the problem-solution oriented frame of experimentation with another frame that is oriented towards the ways in which experiments, collectively and through time, can lead to outcomes of a more systemic nature. We have suggested four processes that lead to the broader embedding of experimental outcomes.

In the academic debate about climate governance, two principal views exist. Hoffmann (2011) has framed 'experimental' climate governance as the multitude of ways of responding to climate change, which function independently of the United Nations negotiations and treaties. Such initiatives operate outside of the frame of multilateral climate governance (van Asselt, Huitema and Jordan, Chapter 2). They involve a variety of actors adopting trial-and-error approaches to problem-solving, operating at and across different scales, and relatively independently of each other. This emerging form of governance has been described as polycentric (Ostrom, 2010; Jordan et al., 2015). This is a different perspective to that taken by Sabel and Zeitlin (2012) and Zeitlin (2015) who talk about the formal European Union (EU) governance architecture as 'experimentalist'. By this they mean a specific system of governance including 'framework' rule and target setting at the EU level, subsidiarity in the implementation of rules and targets to national and local administrations, peer review of target achievement and periodic deliberative review of targets and rules. Sabel and Zeitlin (2012: 6) emphasise the polycentrism and multilevelness of the 'machine for learning from diversity' that constitutes EU rule-making and governance and draw attention to the destabilising effect which the recursive process of peer review between states and revision of rules creates for national administrations. The key differences between these perspectives are that one is global in scope, includes non-state actors and offers a less coordinated bottom-up view of experimentation, while the other is regional in scope, starts with co-ordinated political agreement about high-level goals and operates primarily through the regulatory and administrative procedures of state authorities.

We embrace both perspectives in our discussion about climate governance experiments and their embedding, as this will involve the crafting of formal and informal governance structures to make sense of the realities of climate action on the ground. The search for adequate means to coordinate such governance action – whether this involves the prevalence of top-down, bottom-up or other logics – can be seen as a larger 'experiment' in itself (van Asselt et al., Chapter 2). These differing perspectives also mirror the different forms of embedding that range from scaling up and replication of specific experiments to the aggregation of experimental outcomes in the form of circulation and institutionalisation.

Hoffmann (2011) sees a key role in this experimental governance setting for non-national governments and jurisdictions, including cities, counties, provinces, regions, civil society and corporations. Climate governance experiments have indeed received particular interest in the context of cities and urban environments (e.g. Evans, 2011; Bulkeley et al., 2012; Evans, Karvonen and Raven, 2016) – evidenced also in the contributions of this book (Castán Broto and Bulkeley, Chapter 4; Hölscher et al., Chapter 7; Heiskanen and Matchoss, Chapter 10;

Karvonen, Chapter 11). The urban scale is often seen as a fruitful context for a range of experimentation to take place and connect inhabitants with innovation actors, thus, enabling *experiments in practice* situated in specific urban social, economic, cultural and built milieus. While the importance of national or international scales for climate governance experimentation is acknowledged (Hoffmann, 2011), the potential of experiments has been much less explored on those scales and little literature exists describing experimental governance frameworks beyond REDD+ and the Clean Development Mechanism. This may refer to the haphazard nature of experimentation, possibly making it a poor alternative to national and international regulation, particularly in terms of power and accountability (e.g. Biermann et al., 2012; Hale and Roger, 2014; van Asselt et al., Chapter 2). However, in the absence of effective national and international regulation within international climate policy, and given the complex, contested and uncertain nature of the climate problem, a more experimental approach to governance may be well-fitted to climate governance across different scales. But if this is the case, robust frameworks to evaluate the broader influence and impact of experiments are needed to address the claim that experimentation is a smoke screen for lack of political and practical action to achieve the required transformations in systems and behaviours.

What then might be the features of an 'experimentalist climate governance'? Drawing on the previous discussion, we can identify three primary features. First, climate governance would need to be a frame for experimentation, including the provision of resources, the creation of adequate spaces with exceptional conditions, support for actors and an explicit link between experimental initiatives and wider climate protection and resilience goals. Second, it would be a frame for embedding experimental outcomes, in particular those likely to achieve transformative changes. Taking the example of EU governance, this would include a systematic process for reviewing the outputs, outcomes and potentials of climate governance experiments matched to procedures for embedding promising outputs on a broad scale. Finally, to achieve such broad-scale embedding, experimentalist climate governance would support the four processes of embedding: scaling up, replication, circulation and institutionalisation. There would be an understanding that each of these processes may have a role to play in different settings and that there may be cases in which different processes are employed serially and together to achieve the transformative changes being aimed at.

Such an experimentalist approach seeks to address a central problem for (international) climate governance: the problem of coordination. The Paris Agreement is a radical departure from the idea of the globally co-ordinated, rules-based governance of a global public good problem. It posits instead a loosely coupled, polycentric approach built from actions taken at multiple levels, with regular reporting and peer review. Paris represents an appeal for novel forms of governance beyond the templates offered by globally coordinated, state-led approaches towards solutions beyond, below and outside the state-dominated regime, and towards greater experimentalism. There is a renewed mandate and high-level support for local action (e.g. UN Non-State Actor Zone for Climate Action represents one significant step in this direction) and the strengthening of supporting transnational organisations (e.g. ICLEI-Local Governments for Sustainability, the European Climate-Knowledge Innovation Community [Climate-KIC]) and accounts of a groundswell of local and community initiatives to tackle climate problems.

12.5 Conclusions

In this volume about governance and social experiments aimed at addressing multiple climate change problems, we have proposed to cast a fresh look and focused on their *embedding*. We have sought to focus discussions beyond individual experiments. This new research offers opportunities for further understanding the phenomenon of embedding, new approaches to climate governance and how it can be supported in practice. Such a shift in focus can provide a constructive frame for climate efforts that are increasingly turning towards the role of local and non-state action to deliver urgently required socio-technical transformations (Turnheim et al., Chapter 1; van Asselt et al., Chapter 2). At the same time, it can provide new ideas for state and substate public administrations to think about the role of experiments in their climate policy portfolios. We have extended from the view of Hoffmann (2011) and others in seeing a role for climate governance experiments and beyond also in formal state governance structures – not just at local levels and in informal settings. For both formal and less formal climate governance an urgent requirement for climate governance experimentation is to overcome the current fragmentation of initiatives, and their tendency to remain isolated or short-lived, which ultimately reduces their potential for lasting and wide-ranging change. We suggest that moving the terms of the debate on experimentation towards questions of embedding can constructively shift attention to such strategic and political problems.

Engaging with the embedding of experiments requires the problematisation of the boundary between experiments and their wider outcomes, in scope and time. A number of tensions play out across this boundary: between the particular and generic, the situated and mobile, the isolated and more aggregate, from creative variety to strategic focus, from emergent to institutionalised action and so on. Paying concerted attention to such tensions calls on an ability to facilitate articulation and alignment processes to and fro between the specifics of experimentation-in-context and more generalisable lessons for transformative intervention, recognising the different mechanisms at play and the various relevant dimensions where they play out. This does require the development of new skills and capabilities by governance actors, well beyond the routine management and evaluation of projects. Innovation is something that requires capabilities and investments, and it is risky. Such a focus on embedding should not deflect attention away from the specifics of local projects and how they can be carried out and supported, but rather attract attention towards an equally important aspect related to the wider relevance and positive impact of experimentation for climate governance.

The contributions brought together in this volume have considered a wide range of views on how experiments become embedded, actively mobilising the disciplinary richness associated with the emerging phenomenon of climate governance experimentation. These views go well beyond a conventional focus on scaling and linear diffusion models. Alternatives to scaling considered here include the replication of experiments in new contexts, the circulation of experimental outcomes in different forms (knowledge, people, blueprints and so on) and the transformation of structures and governance regimes as experimental outcomes force a reconsideration of core rules and institutional logics.

Embedding, as a process, is difficult. It requires active work and engagement by participants, as well as time for systemic, institutional and actor-based dynamics to play out. Articulation and alignment are the main master processes at work in embedding, played out at different levels and across institutional, technical, social and political dimensions. Embedding is mediated through contexts, as well as by the specific politics of experimentation practices, including learning and knowledge accumulation, the build-up of networks and coalitions of actors, the emergence of new roles (e.g. intermediation, facilitation and governance entrepreneurship) and mobilisation around shared expectations and collective narratives.

The embedding of climate governance experiments concerns experiments themselves, as well as the outputs and outcomes of experimentation. In the former, central questions relate to how experiments can be reframed or repurposed to better 'fit' existing regimes and a variety of contexts, and how regimes can learn to handle, manage or orchestrate experimentation. In the latter, processes of transformative change become the main focus, for which individual experiments may be an agent or accessory of more structural outcomes. Furthermore, and particularly when considering the evaluation of experiments and their outputs and outcomes, it becomes necessary to look beyond the immediate outputs (intended and unintended) of experiments and to engage with the processes that enable experimentation and its embedding.

To this, we wish to add a reminder of the potentially empowering value of experimentation. Genuine experiments can be seen as spaces of emancipation wherein new actors (often excluded from currently prevailing arrangements) can become empowered to contribute to collective problem-solving efforts. Recognising this emancipatory character of experimentation further challenges common views on scaling, diffusion and embedding as requiring the extraction and purification of situated knowledge and may justify a search for climate governance that respects and draws on the desires of local actors not to see their initiatives uprooted or co-opted, but simply acknowledged as relevant and meaningful. It remains to be seen whether such empowerment and emancipation pathways can become aligned with notions of 'orchestration' increasingly put forward within climate governance debates as means to recognise the relevance of more transnational and local action without imposing top-down logics (Abbott, 2012; Hale and Roger, 2014; Chan et al., 2015).

References

Abbott, K. W. (2012). The transnational regime complex for climate change. *Environment and Planning C: Government and Policy*, 30(4), 571–590. doi: 10.1068/c11127.

Abernathy, W. J., and Utterback, J. M. (1978). Patterns of industrial innovation. *Technology Review*, 80(7), 40–47.

Berkhout, F., Verbong, G., Wieczorek, A., Raven, R., Lebel, L., and Bai, X. (2010). Sustainability experiments in Asia: Innovations shaping alternative development pathways? *Environmental Science & Policy*, 13(4), 261–271.

Biermann, F., Abbott, K., Adresen, S., Bäckstrand, K., Bernstein, S., Betsill, M.M., Bulkeley, H., Cashore, B., Clapp, J., Folke, C., Gupta, A., Gupta, J., Haas, P.M., Jordan, A., Kanie, N., Kluvánková-Oravská, T., Lebel, L., Liverman, D., Meadowcroft, J., Mitchell, R.B., Newell, P., Oberthür, S., Olsson, L., Pattberg, P., Sánchez-Rodríguez, R., Schroeder, H., Underdal, A., Camargo Vieira, S., Vogel, C., Young, O.R., Brock, A. and Zondervan R. (2012). Transforming governance and institutions for global sustainability: Key insights from the Earth System Governance Project. *Current Opinion in Environmental Sustainability*, 4(1), 51–60.

Bos, J. J., and Brown, R. R. (2012). Governance experimentation and factors of success in socio-technical transitions in the urban water sector. *Technological Forecasting and Social Change*, 79(7), 1340–1353. doi: 10.1016/j.techfore.2012.04.006.

Brown, H. S., and Vergragt, P. J. (2008). Bounded socio-technical experiments as agents of systemic change: The case of a zero-energy residential building. *Technological Forecasting and Social Change*, 75(1), 107–130.

Bulkeley, H., Andonova, L, Bäckstrand, K., Betsill, M., Compagnon, D., Duffy, R., Kolk, A., Hoffmann, M., Levy, D., Newell, P., Milledge, T., Paterson, M., Pattberg, P. and vanDeveer, S. (2012). Governing climate change transnationally: Assessing the evidence from a database of sixty initiatives. *Environment and Planning C: Government and Policy*, 30(4), 591–612.

Cabinet Office (2003). Trying it out: The role of 'pilots' in policy-making. In *Report of a Review of Government Pilots*.

Chan, S., van Asselt, H., Hale, T., Abbott, K.W., Beisheim, M., Hoffmann, M., Guy, B., Hohne, N., Hsu, A., Pattberg, P., Pauw, P., Ramstein, C. and Widerberg, O. (2015). Reinvigorating international climate policy: A comprehensive framework for effective nonstate action. *Global Policy*, 6(4), 466–473.

Evans, J., Karvonen, A., and Raven, R. (eds.) (2016). *The Experimental City*. Abingdon: Routledge.

Evans, J. P. (2011). Resilience, ecology and adaptation in the experimental city. *Transactions of the institute of British Geographers*, 36(2), 223–237.

Geels, F., and Raven, R. (2006). Non-linearity and expectations in niche-development trajectories: Ups and downs in Dutch Biogas Development (1973–2003). *Technology Analysis & Strategic Management*, 18(3–4), 375–392. doi: 10.1080/ 09537320600777143.

Geels, F. W., and Deuten, J. J. (2006). Local and global dynamics in technological development: A socio-cognitive perspective on knowledge flows and lessons from reinforced concrete. *Science and Public Policy*, 33(4), 265–275.

Grin, J., Rotmans, J., and Schot, J. (2010). *Transitions to Sustainable Development: New Directions in the Study of Long Term Transformative Change*. London: Routledge.

Hale, T., and Roger, C. (2014). Orchestration and transnational climate governance. *Review of International Organizations*, 9(1), 59–82.

Hartmann, A., and Linn, J. (2009). Scaling up: A path to effective development. In von Braun, J., Vargas Hill, R., and Pandya-Lorch, R. (eds.), *The Poorest and Hungry: Assessments, Analyses, and Actions*. International Food Policy Research Institute, 449–558.

Hoffmann, M. J. (2011). *Climate Governance at the Crossroads: Experimenting with a Global Response after Kyoto*. Oxford: Oxford University Press.

Jolly, S., Raven, R., and Romijn, H. (2012). Upscaling of business model experiments in off-grid PV solar energy in India. *Sustainability Science*, 7(2), 199–212. doi: 10.1007/ s11625-012–0163-7.

Jordan, A., and Huitema, D. (2014). Policy innovation in a changing climate: Sources, patterns and effects. *Global Environmental Change*, 29, 387–394.

Jordan, A. J., Huitema, D., Hildén, M., van Asselt, H., Rayner, T.J., Schoenefeld, J.J., Tosun, J., Forster, J. and Boasson, E.L. (2015). Emergence of polycentric climate governance and its future prospects. *Nature Climate Change*, 5(11), 977–982. doi: 10.1038/nclimate2725.

Kemp, R. P. M., Rip, A., and Schot, J. W. (2001). Constructing transition paths through the management of niches. In Garud, R., and Karnoe, P. (eds.), *Path Dependence and Creation*. Mahwah, NJ: Lawrence Erlbaum, 269–299.

Kivimaa, P. (2014). Government-affiliated intermediary organisations as actors in system-level transitions. *Research Policy*, 43, 1370–1380.

Kivimaa, P., Hildén, M., Huitema, D., Jordan, A., Newig, J. (2017). Experiments in climate governance: A systematic review of research on energy and built environment transitions. *Journal of Cleaner Production*, 169: 17–29.

Kivimaa, P., and Kern, F. (2016). Creative destruction or mere niche support? Innovation policy mixes for sustainability transitions. *Research Policy*, 45(1), 205–217.

Laakso, S., Berg, A., and Annala, M. (2017). Dynamics of experimental governance: A meta-study of functions and uses of climate governance experiments. *Journal of Cleaner Production*, 1–9. doi: 10.1016/j.jclepro.2017.04.140.

Nair, S., and Howlett, M. (2016). Meaning and power in the design and development of policy experiments. *Futures*, 76, 67–74. doi: 10.1016/j.futures.2015.02.008.

Ostrom, E. (2010). Beyond markets and states: Polycentric governance of complex economic systems. *American Economic Review*, 100(3), 641–672. doi: 10.1257/ aer.100.3.641.

Primmer, E., Paloniemi, R., Simila, J. and Barton, D. (2013). Evolution in Finland's Forest Biodiversity Conservation payments and the institutional constraints on establishing new policy. *Society & Natural Resources*, 26(10), 1137–1154. doi: 10.1080/08941920.2013.820814.

Raven, R., Van Den Bosch, S., and Weterings, R. (2010). Transitions and strategic niche management: Towards a competence kit for practitioners. *International Journal of Technology Management*, 51(1), 57–74. doi: 10.1504/IJTM.2010.033128.

Sabel, C. F., and Zeitlin, J. (2012). Experimentalist governance. In Levi-Faur, D. (ed.), *The Oxford Handbook of Governance*. Oxford: Oxford University Press.

Smith, A., and Raven, R. (2012). What is protective space? Reconsidering niches in transitions to sustainability. *Research Policy*, 41(6), 1025–1036. doi: 10.1016/j.respol.2011.12.012.

Tassey, G. (2014). Innovation in innovation policy management: The Experimental Technology Incentives Program and the policy experiment. *Science and Public Policy*, 41(4), 419–424.

Turnheim, B., and Geels, F. W. (2012). Regime destabilisation as the flipside of energy transitions: Lessons from the history of the British coal industry (1913–1997). *Energy Policy*, 50, 35–49.

Upham, P., Kivimaa, P., Mickwitz, P., Astrand, K. (2014). Climate policy innovation: A sociotechnical transitions perspective. *Environmental Politics*, 23(5), 774–794.

Zeitlin, J. (2015). *Extending Experimentalist Governance? The European Union and Transnational Regulation*. Oxford: Oxford University Press.

Index